Electronic Power Supplies

HOLT, RINEHART AND WINSTON SERIES IN ELECTRONICS TECHNOLOGY

A COMPOSITE INTRODUCTION TO ELECTRIC CIRCUITS
by George Caliva

BASIC ELECTRON DEVICES *by E. George Griffith*

VOLTAGE AND POWER AMPLIFIERS *by Robert E. Sentz*

SPECIAL ELECTRON DEVICES *by Walter Sowa and James Toole*

TEST EQUIPMENT AND TROUBLESHOOTING PROCEDURES
by George Caliva

ELECTRONIC POWER SUPPLIES *by Joseph Grabinski*

PULSE AND DIGITAL CIRCUITS *by John Canavan*

FEEDBACK AMPLIFIERS AND OSCILLATORS
by Robert E. Sentz and Robert A. Bartkowiak

INFORMATION TRANSMITTAL AND COMMUNICATING SYSTEMS
by John Paul Froehlich

TRANSMISSION LINES AND ANTENNAS *by Stephen Roman*

ROTATING MACHINERY *by Harold J. Herbein*

ALGEBRAIC TRANSIENT ANALYSIS *by Robert Miller*

TECHNICAL CALCULUS *by Dennis H. Price and H. Richard Stillwell*

Electronic Power Supplies

JOSEPH GRABINSKI

Hartford State Technical College

UNDER THE EDITORSHIP OF JOHN M. DOYLE

HOLT, RINEHART AND WINSTON, INC.

NEW YORK, CHICAGO, SAN FRANCISCO, ATLANTA, DALLAS
MONTREAL, TORONTO, LONDON

Library of Congress Catalog Card Number 68-25813

SBN: 03-071370-6

PRINTED IN THE UNITED STATES OF AMERICA

1 2 3 4 5 6 7 8 9

Illustrations by Visual Direction

To My Parents
With Love and Appreciation

Preface

This book is one volume in the ELECTRONICS TECHNOLOGY SERIES designed for use in technical institutes and in both junior and community colleges. It will also be useful as a text in industrial training programs for upgrading currently employed technicians and in independent self-study programs.

The principal objective is to provide a clear understanding of power supply circuits and the waveforms associated with these circuits.

Assumed prerequisites are: (1) A course in d-c and a-c circuits; (2) an understanding of the basic operation of solid-state and vacuum diodes; and (3) a working knowledge of algebra, trigonometry, and elementary calculus.

The book is written in a clear, concise manner with special attention given to those subjects which are normally most difficult to understand. All mathematical developments are presented without skipping steps—a traditional source of irritation to the reader who is trying to follow the mathematical development of a topic.

Waveforms for all circuits are discussed and illustrated extensively. Not only do they help the reader to visualize the physical operation of the circuit, but they serve as an aid in the reading of the text itself.

Some of the outstanding features of this book are: (1) A thorough discussion of three-phase rectification; (2) the magnetic amplifier—a too-often neglected component of power supplies; and (3) a full treatment of regulation circuits that incorporate silicon-controlled rectifiers, zener diodes, thyratrons, and voltage-regulator tubes.

Comprehensive problem and question sections appear at the end of each chapter. Solutions are available to teachers.

Joseph Grabinski

Hartford, Conn.
September 1967

Contents

Electronic Power Supplies

Single-Phase Rectifier Circuits

1

Most commercially available power is of the a-c variety. Unfortunately, most electronics equipment requires d-c power for proper operation. Some means must be employed, therefore, to convert the a-c power to d-c power. A device used for this purpose is called a *rectifier*. When a rectifier is used with required components, the assembly is known collectively as a *rectifier circuit*. When certain types of other components are added to a rectifier circuit, for purposes to be discussed later, the assembly is usually called a *power supply*.

This chapter is concerned only with rectifier circuits which operate from a single-phase, a-c power input.

1-1 THE BASIC RECTIFIER

The basic characteristic of a rectifier is that it is a unilateral device; that is, it will pass current in one direction only. A mechanical analogy of the rectifier is illustrated in Fig. 1-1. Here, a simple flapper-type valve is inserted into a pipeline. When pressure is exerted in the direction shown in Fig. 1-1(a), the valve moves backward or opens, permitting passage or flow. When pressure is exerted in the opposite direction, however, as shown in Fig. 1-1(b), the valve is forced against the stop, this closure stopping the flow.

In a simple rectifier of the type found in electric circuits, shown in Fig. 1-2(a), 110 v ac are applied across a transformer. In series with the transformer secondary are a *diode* and a load resistor. The symbol I_L is used to indicate load current, and V_L indicates the voltage drop appearing across the load resistor. With reference to the diode, the arrowhead represents the *anode* (plate) and the straight line represents the *cathode*.

An *ideal diode* is a fictitious device that is assumed to exhibit zero resistance when its anode becomes positive with respect to its cathode, and only under these conditions is the ideal diode assumed to conduct (pass current). Whenever the anode is negative with respect to the cathode, the ideal diode is assumed to possess infinite resistance, and current through the device is zero. Although no practical diode exhibits these theoretical characteristics, ideal conditions are assumed for the sake of simplicity. No appreciable error results from this assumption.

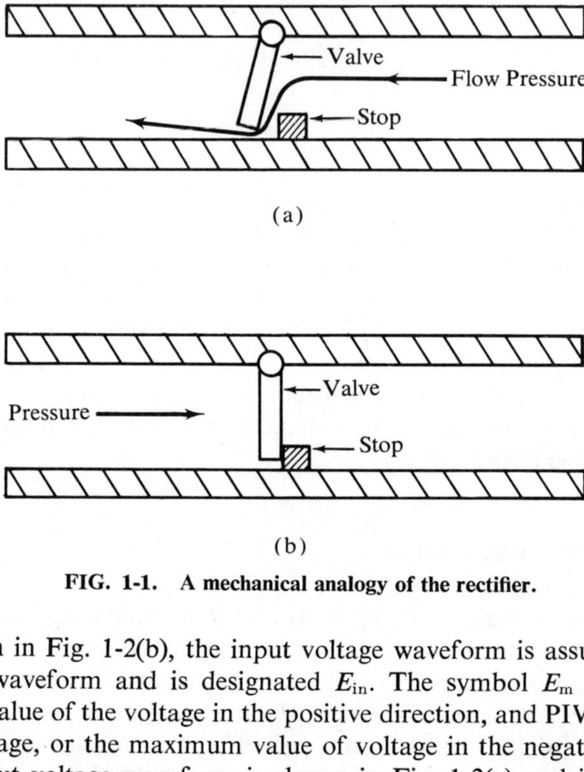

FIG. 1-1. A mechanical analogy of the rectifier.

As shown in Fig. 1-2(b), the input voltage waveform is assumed to be a sinusoidal waveform and is designated E_{in}. The symbol E_m refers to the maximum value of the voltage in the positive direction, and PIV to the peak-inverse voltage, or the maximum value of voltage in the negative direction.

The output voltage waveform is shown in Fig. 1-2(c) and is termed V_L, since it is the voltage appearing across the load resistor. The symbol V_p represents the peak value of V_L, and V_{av} and V_{rms} represent the average and rms values, respectively.

By referring to the three parts of Fig. 1-2, it can be seen that when E_{in} causes the voltage at the top of the transformer secondary to be positive (+), the anode of the diode is positive with respect to its cathode; consequently, the diode conducts, current passes through the load resistor, and a voltage appears across the load resistor. On the other hand, when E_{in} causes voltage at the top of the transformer secondary to be negative, the anode of the diode is negative with respect to its cathode; therefore, the diode cannot

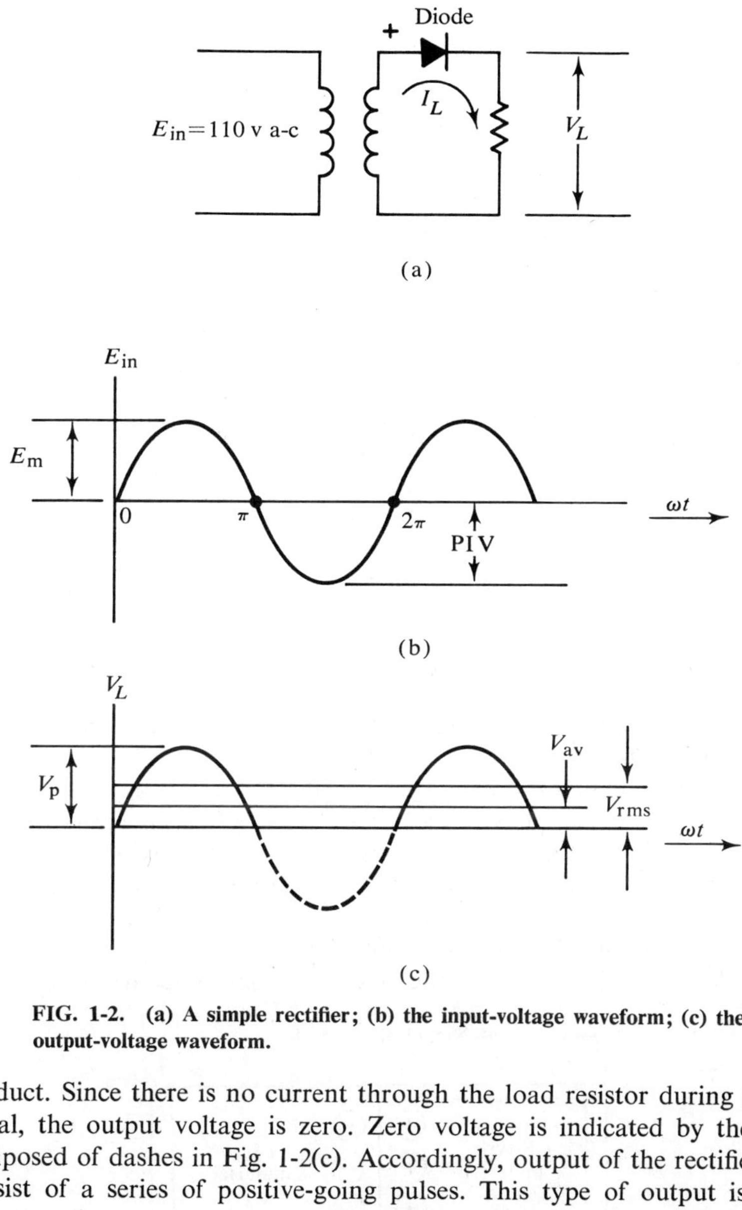

FIG. 1-2. (a) A simple rectifier; (b) the input-voltage waveform; (c) the output-voltage waveform.

conduct. Since there is no current through the load resistor during this interval, the output voltage is zero. Zero voltage is indicated by the curve composed of dashes in Fig. 1-2(c). Accordingly, output of the rectifier must consist of a series of positive-going pulses. This type of output is called *pulsating* direct current.

Before leaving Fig. 1-2, note that conventional current is indicated, that is, the current flows from positive to negative. When the concept of electron flow is used, current would be shown as moving from negative to positive. Since the choice is arbitrary, it should cause no confusion.

1-2 COMMON RECTIFIER ELEMENTS

A rectifier element known as a *selenium cell*, shown in Fig. 1-3, is generally used in applications where low ambient heating is experienced and where

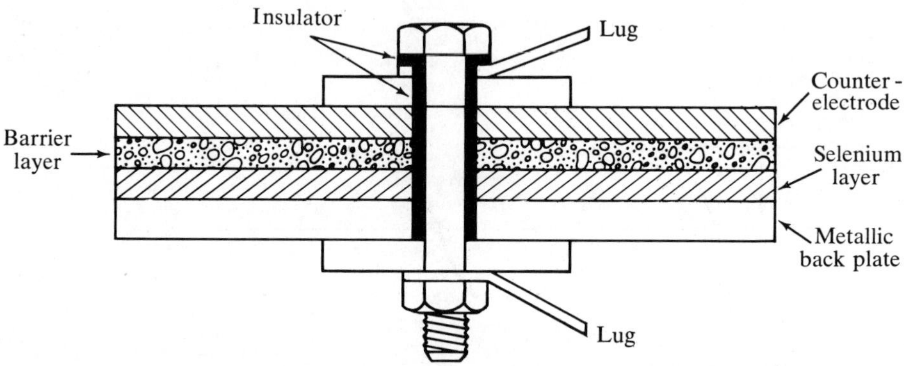

FIG. 1-3. Cross-sectional view of a selenium cell.

physical size is not a design problem. After a specially treated film of selenium has been deposited in a series of heat treatments on the cell's backing plate of iron or aluminum, a low-melting-point alloy is sprayed onto the selenium surface. This layer of alloy is called the *counterelectrode*. By an electrochemical process a blocking layer (barrier) is formed between the selenium surface and the counterelectrode. Selenium plates can block d-c voltages as high as 75 v per plate and can operate at current densities of approximately 70 mA/cm^2, with an efficiency of about 75 percent.

The *silicon* diode, shown in Fig. 1-4, is a semiconductor device that is

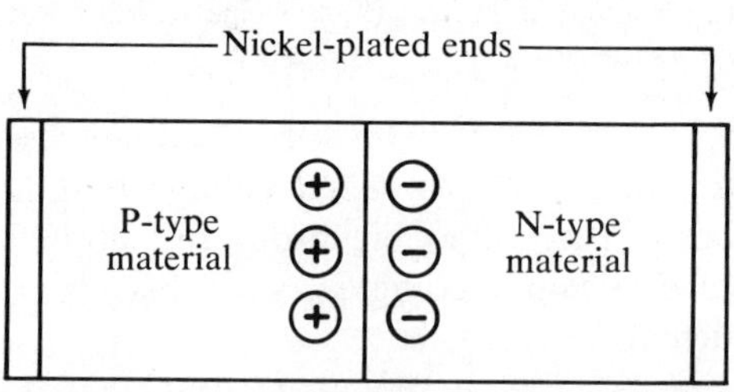

FIG. 1-4. Pictorial representation of a silicon diode.

often used as a rectifier. It is produced by doping (adding impurities to) a silicon crystal in the same manner as that used to produce a transistor crystal. The silicon rectifier crystal is made by the diffusion of phosphorus into one side of a silicon crystal and of boron into the other side. Nickel-plating each side of the crystal makes possible the attachment of leads for connection into an external circuit.

The voltage drop with a silicon diode during conduction is lower than that of a selenium unit. Silicon rectifiers also exhibit better reverse characteristics than those of selenium units. Maximum reverse voltage may rise above 200 volts at a reverse current of less than 10^{-6} A. Silicon rectifiers are made in single units able to handle as many as 100 A.

1-3 THE HALF-WAVE RECTIFIER WITH A RESISTIVE LOAD

Referring to the waveforms shown in Fig. 1-2(b) and (c), note the following:

1. The peak values of the d-c voltage and current
2. The peak-inverse voltage applied across the rectifying element when it is nonconducting
3. The average values of d-c voltage and current
4. The rms values of voltage and current
5. The average load power
6. The magnitude of pulsations (termed *ripple*) appearing across the load

If the rectifying element had no forward voltage-drop, the voltage across the load resistor during periods of conduction would, of course, be equal to the input voltage at every instant, and the peak-load voltage would be equal to the peak-input voltage. As noted previously, it is common practice to assume that there is no voltage drop in the forward direction, that is, when the rectifier is conducting, so that $V_L = E_{in}$.

The peak-inverse voltage (also called the reverse voltage) across a half-wave rectifier with a resistive load is the peak value of the voltage wave. This is the maximum-reverse voltage impressed across the rectifier when it is not conducting.

The average values of the voltage and current are the respective averages of the instantaneous values during a given period of time. Figure 1-5 shows the rectified output voltage for one and one-half cycles of input voltage. The rectifier conducts for only one-half of the input cycle, but the average value of the voltage wave applies to the entire cycle. To find the average value, therefore, first find the average of all instantaneous values of voltage,

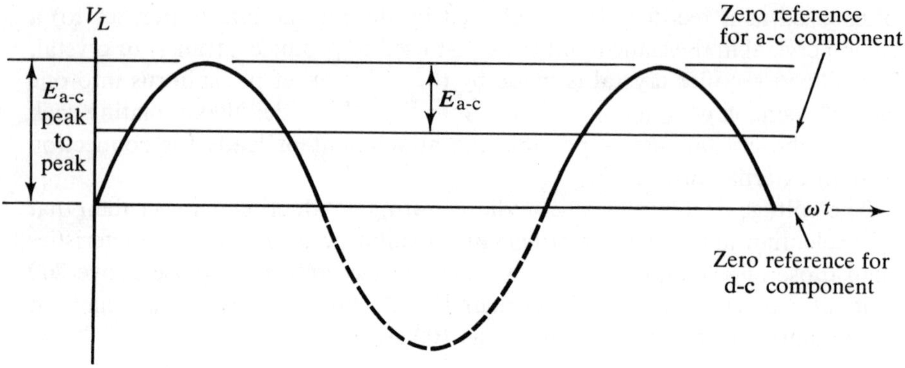

FIG. 1-5. Load voltage from a half-wave rectifier.

termed e, during the half-cycle of conduction and then divide this average by the period of one cycle.

Summing all instantaneous values of e during a half-cycle,

$$\sum_{o}^{\pi} e = \int_{o}^{\pi} E_{\mathrm{m}} \sin \omega t(d\omega t) \tag{1-1}$$

and the average value for a full cycle is

$$E_{\mathrm{av}} = \frac{1}{2\pi} \int_{0}^{\pi} E_{\mathrm{m}} \sin \omega t(d\omega t) \tag{1-2}$$

Eq. 1-2 reduces to

$$E_{\mathrm{av}} = \frac{E_{\mathrm{m}}}{2\pi} \Big[-\cos \omega t \Big]_{0}^{\pi} \tag{1-3}$$

and may be rewritten as

$$E_{\mathrm{av}} = \frac{E_{\mathrm{m}}}{2\pi} [-\cos \pi - (-\cos 0)] \tag{1-4}$$

A final simplification of Eq. 1-4 gives

$$E_{\mathrm{av}} = \frac{E_{\mathrm{m}}}{\pi} \tag{1-5}$$

The current and voltage in a half-wave rectifier have the same waveform; therefore,

$$I_{\mathrm{av}} = \frac{E_{\mathrm{m}}}{\pi R} \tag{1-6}$$

or

$$I_{\mathrm{av}} = \frac{I_{\mathrm{m}}}{\pi} \tag{1-7}$$

The effective (rms) values of voltage and current are defined in terms of their ability to deliver power to a resistance load. The instantaneous power in any circuit can be expressed as

$$p = i^2 R \tag{1-8}$$

where p and i represent the instantaneous values of power and current, respectively. For the half-wave circuit, the instantaneous current is

$$i = I_m \sin \omega t \qquad 0 \leqq \omega t \leqq \pi \tag{1-9}$$

The instantaneous power then becomes

$$p = (I_m \sin \omega t)^2 R \qquad 0 \leqq \omega t \leqq \pi \tag{1-10}$$

The average power over a full cycle is the sum of all instantaneous values of power divided by the period of one full cycle.

$$P_{av} = \frac{1}{2\pi} \int_0^{\pi} R(I_m \sin \omega t)^2 \, d\omega t$$

which may be rewritten as

$$P_{av} = \frac{R(I_m)^2}{2\pi} \left[\frac{\omega t}{2} - \sin 2\,\omega t \right]_0^{\pi} \tag{1-11}$$

and reduces to

$$P_{av} = R\,\frac{(I_m)^2}{4} \tag{1-12}$$

The average power in any circuit is defined as

$$P_{av} = (I_{rms})^2 R \tag{1-13}$$

From Eq. 1-12, the rms current from a half-wave rectifier must be

$$(I_{rms})^2 = \frac{(I_m)^2}{4}$$

or

$$I_{rms} = \frac{I_m}{2} \tag{1-14}$$

Because the voltage and current in a resistive circuit have the same waveform, the rms voltage across the load is

$$E_{rms} = \frac{E_m}{2} \tag{1-15}$$

If the rectifying element were removed from a circuit, the a-c power delivered to the load would be

$$P_{a\text{-}c} = \left(\frac{I_m}{\sqrt{2}} \right)^2 R$$

or

$$P_{a\text{-}c} = \frac{(I_m)^2}{2} R$$

Without the rectifying element in the circuit, twice as much power would be delivered as that with the rectifier. However, half the available power is not wasted. Since current during one-half of the cycle is essentially zero, the power lost during this portion of the cycle is negligible. If the rectifying element has low forward resistance and low reverse current, the power loss in the circuit is very low when the diode is not conducting, that is, when the diode is reverse biased.

1-4 RIPPLE FACTOR

The output voltage wave of a rectifier consists of a d-c component with an a-c component, called *ripple*, superimposed. The ratio of the effective (rms) value of a-c ripple to the average value of the d-c component is called the *ripple factor* and is a measure of the purity of rectifier output. Mathematically,

$$\text{Ripple factor} = \frac{\text{effective (rms) value of ac}}{\text{average value of dc}} \tag{1-16}$$

The rms value of the combined a-c and d-c components is expressed as

$$E_{rms} = \sqrt{E_{av}^2 + E_{a\text{-}c}^2} \tag{1-17}$$

Where E_{rms} = effective value of the combined a-c and d-c components
E_{av} = average value of the d-c component
$E_{a\text{-}c}$ = effective (rms) value of the a-c ripple

Rearranging Eq. 1-17 to solve for $E_{a\text{-}c}$

$$E_{a\text{-}c} = \sqrt{E_{rms}^2 - E_{av}^2} \tag{1-18}$$

Combining Eqs. 1-16 and 1-18, the ripple factor, designated by the Greek letter gamma, γ, is

$$\gamma = \frac{\sqrt{E_{rms}^2 - E_{av}^2}}{E_{av}} \tag{1-19}$$

By rearranging Eq. 1-19, γ may be expressed as

$$\gamma = \sqrt{\left(\frac{E_{rms}}{E_{av}}\right)^2 - 1} \tag{1-20}$$

The *form factor F* of any wave is defined as the ratio of the effective value to the average value of the wave, so that ripple factor may be expressed as $\gamma = \sqrt{F^2 - 1}$.

From Eq. 1-18 it can be seen that if an average-reading d-c voltmeter and an a-c rms voltmeter are connected to a load, the readings obtained can be used to calculate the ripple factor of the rectified voltage.

Example 1-1 The load voltage of a rectifier circuit is found to consist of an average d-c voltage of 80 v and an a-c voltage with an effective value of 20 v. For the output waveform, compute the: (1) ripple factor, (2) value of the combined a-c and d-c components of voltage, and (3) form factor of the wave.

1. From Eq. 1-16, ripple factor $\frac{E_{\text{a-c}}}{E_{\text{av}}}$

$$\gamma = \frac{20}{80} = 0.25, \text{ or } 25 \text{ percent}$$

2. From Eq. 1-17

$$E_{\text{rms}} = \sqrt{(E_{\text{av}})^2 + (E_{\text{a-c}})^2}$$

$$E_{\text{rms}} = \sqrt{(80)^2 + (20)^2}$$

$$E_{\text{rms}} = 82.5 \text{ v}$$

3. By definition, the form factor is the ratio of the rms voltage to the average voltage. Therefore,

$$F = \frac{82.5}{80} = 1.03$$

1-5 THE HALF-WAVE RECTIFIER WITH A CAPACITIVE LOAD

Rectifying circuits may be called upon to deliver power to reactive (capacitive and/or inductive) as well as to resistive loads; therefore, it is of interest to examine the behavior of a rectifier circuit in supplying reactive loads.

When a capacitor is connected to a rectifying element, as shown in Fig. 1-6(a), the resulting voltage, E_C, and current, I_C, waveforms are as shown in Fig. 1-6(b). Assume, for analytical purposes, that the input voltage is applied to the circuit when the value of the voltage wave is zero and about to swing in the positive direction; assume in addition that the rectifier and capacitor are perfect (ideal). The supply voltage appears directly across the capacitor at time $t = 0$. If the input voltage is described as $e_{\text{in}} = E_{\text{m}} \sin \omega t$, then, since current leads voltage in a capacitive circuit, the current can be expressed as $i_C = I_{\text{m}} \cos \omega t$.

In the assumed circuit current rises instantaneously to its maximum value,

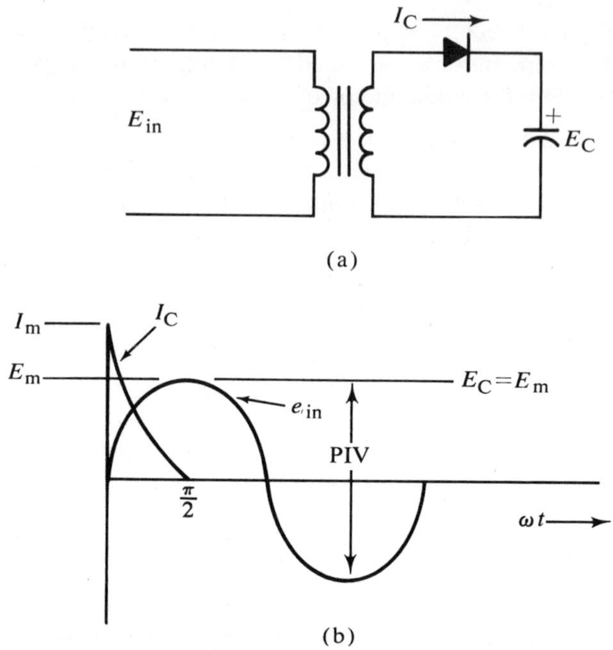

FIG. 1-6. (a) Half-wave rectifier with capacitive load; (b) associated current- and voltage-waveforms.

decaying to zero as the capacitor charges. When the input voltage is at $\pi/2$, the capacitor is charged to E_m with the polarity shown, and the current is zero. The voltage across the capacitor follows the input voltage exactly during the first one-quarter cycle while the capacitor is charging. But, as the supply voltage decreases, the charge on the capacitor must decrease if the capacitor voltage is to equal the supply voltage. In order for the charge to decrease, current must pass from the positive terminal of the capacitor back through the rectifier. Since the rectifier will not allow conduction in this direction, the current remains at zero, and the voltage and charge on the capacitor remain constant. Current in the circuit is zero, and no change in E_C takes place after the first one-quarter cycle. The voltage E_C remains at E_m with a positive polarity, as shown. Thus, when the input voltage is at $-E_m$, the peak-inverse voltage across the diode is 2 E_m.

1-6 THE HALF-WAVE RECTIFIER WITH AN *RC* LOAD

When a capacitor is connected in parallel with a resistive load, as shown in Fig. 1-7, the operating conditions are considerably different from those

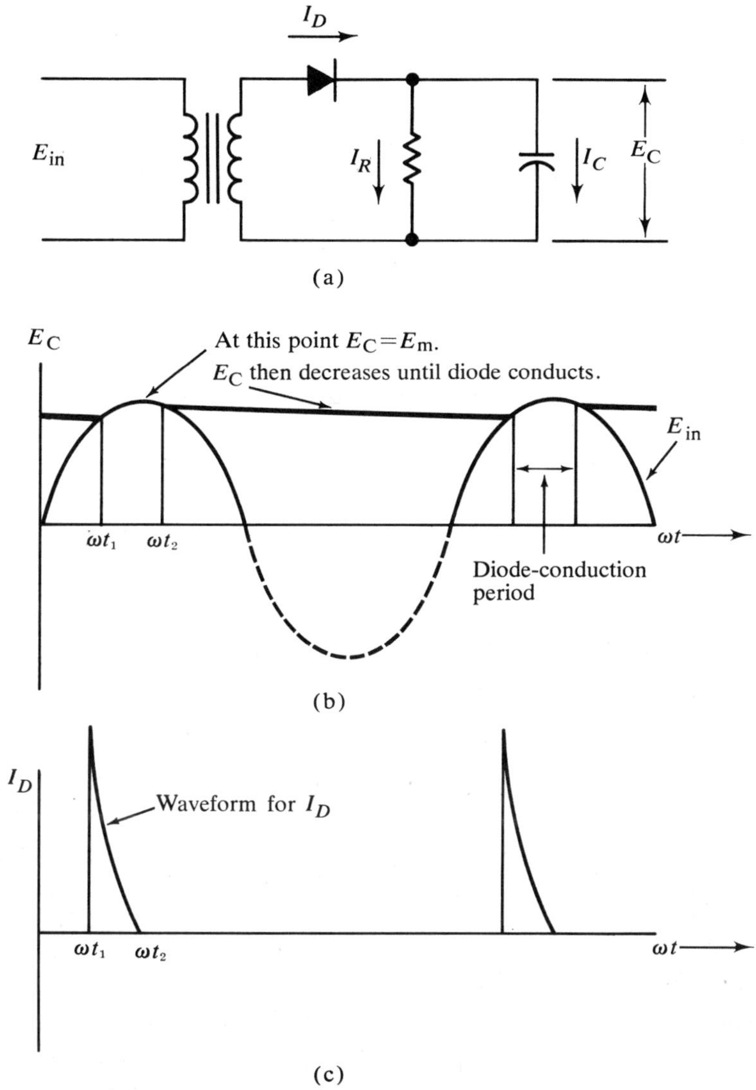

FIG. 1-7. (a) Half-wave rectifier with *RC* load; (b) Load voltage with *RC* load; (c) Diode current with *RC* load.

in a rectifier circuit having a purely resistive load. During the initial charging time the capacitor voltage, E_C, reaches a value equal to the maximum input voltage, and a charge, designated by the letter symbol Q, is stored by the capacitor. When the supply voltage falls, the capacitor discharges through the load resistor. Current is, therefore, maintained in the load during the period when the input voltage is falling to zero and also during the negative half-cycle as shown in Fig. 1-7(b). Because current is maintained through the resistor when the diode is not conducting, the average d-c value is increased, and the ripple factor is decreased.

The diode conducts only during the time when the input voltage is greater than the capacitor voltage. During this conduction time the diode must supply current to the load and also supply the capacitor with the charge necessary to maintain the increase in average current to the load. If this higher average load current is to be maintained when the diode is not conducting, the diode current I_D must be very high during the conduction period. The waveform for the current in the diode is shown in Fig. 1-7(c). The Kirchhoff equation for the currents in Fig. 1-7 when the diode is conducting is

$$I_D = I_R + I_C$$

The instantaneous current relationships may be written as

$$i_D = \frac{E_m}{R} \sin \omega t + \frac{E_m}{X_C} \cos \omega t$$

or

$$i_D = \frac{E_m}{R} \sin \omega t + \omega C E_m \cos \omega t \tag{1-21}$$

At ωt_2 the diode ceases to conduct and $i_D = 0$. At ωt_2, therefore,

$$-i_C = i_R \tag{1-22}$$

This may be expressed as

$$-\omega C E_m \cos \omega t_2 = \frac{E_m}{R} \sin \omega t_2 \tag{1-23}$$

Rearranging Eq. 1-23,

$$\frac{\sin \omega t_2}{\cos \omega t_2} = -\omega RC \tag{1-24}$$

from which

$$\tan \omega t_2 = -\omega RC \tag{1-25}$$

The angle at which the diode ceases to conduct is ωt_2, and

$$\omega t_2 = \tan^{-1} -\omega RC \tag{1-26}$$

The minus sign indicates that this angle is in the second quadrant; as ωRC increases, the angle approaches 90°.

Referring to Eqs. 1-21 and 1-23 it can be seen that the current in the capacitor is positive during the time the diode conducts and negative when

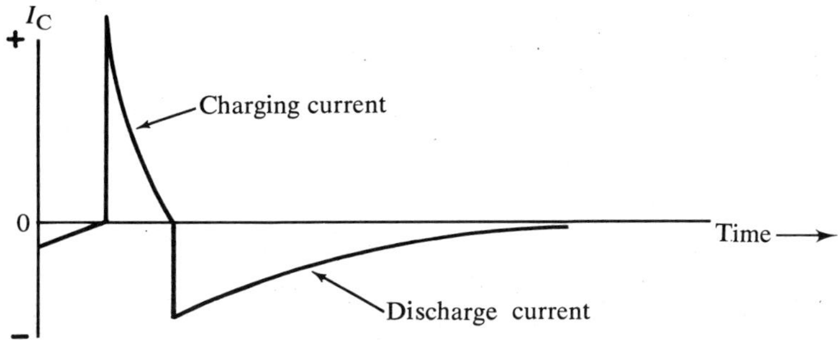

(a)

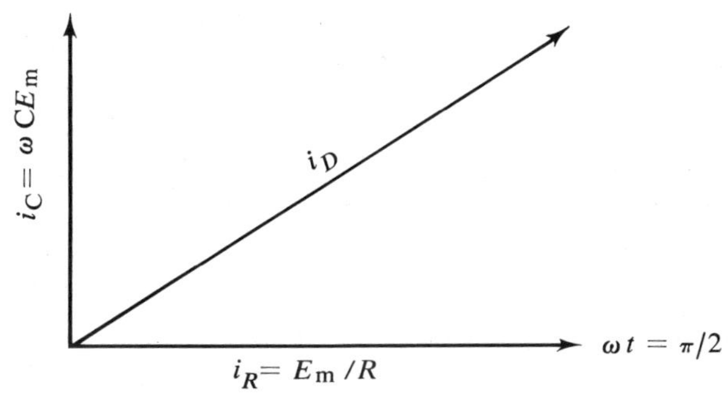

(b)

FIG. 1-8. (a) Waveform of the capacitor charging and discharging current; (b) Vector diagram of capacitor and load currents.

the diode current is zero. From these equations it is obvious that the capacitor current reverses, and that the capacitor supplies current to the load when the diode is cut off. The waveform for the capacitor current is shown in Fig. 1-8(a).

Equation 1-21 states that during the diode conduction time the total diode current is composed of both an in-phase component, $i_R = (E_m/R)\sin \omega t$, and a leading component, $i_C = \omega C E_m \cos \omega t$. The vector diagram for summing these currents when $E_{in} = E_m$ is shown in Fig. 1-8(b). From the vector diagram

$$i_D = \sqrt{(\omega C E_m)^2 + \left(\frac{E_m}{R}\right)^2}$$

or

$$i_D = E_m \sqrt{\frac{R^2C^2\omega^2 + 1}{R^2}}$$

Rearranging,

$$i_D = \frac{E_m}{R}\sqrt{1+\omega^2R^2C^2} \tag{1-27}$$

Equation 1-27 states that the diode current is equal to E_m/R when C is zero, and that the instantaneous current increases as the RC time constant increases. This current also increases as the frequency of the source increases.

To calculate the average d-c voltage from a half-wave rectifier with an RC load, certain assumptions must be made to simplify the mathematical procedure; namely, (a) the circuit contains a perfect rectifier and a perfect capacitor, (b) the capacitor charges instantaneously when the supply voltage is greater than the capacitor voltage, and (c) the capacitor discharges at a linear rate through the resistor. See Fig. 1-9.

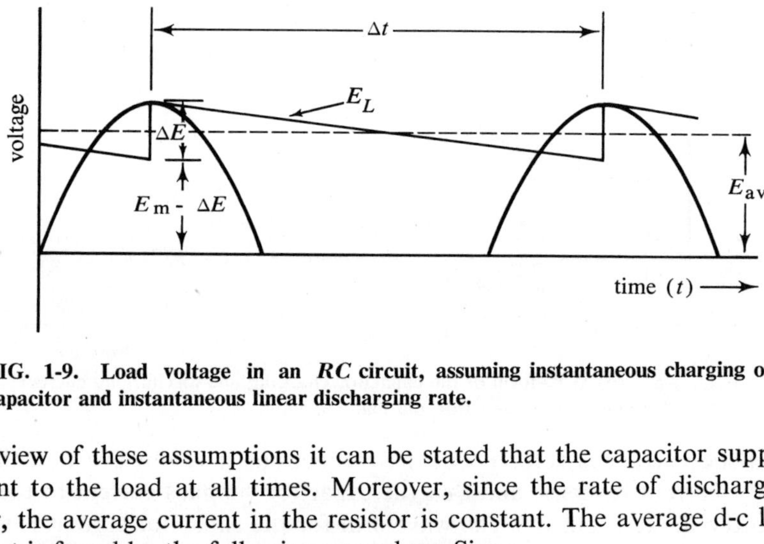

FIG. 1-9. Load voltage in an *RC* circuit, assuming instantaneous charging of capacitor and instantaneous linear discharging rate.

In view of these assumptions it can be stated that the capacitor supplies current to the load at all times. Moreover, since the rate of discharge is linear, the average current in the resistor is constant. The average d-c load current is found by the following procedure. Since

$$\Delta Q = C\,\Delta E$$

and

$$\frac{\Delta Q}{\Delta T} = C\frac{\Delta E}{\Delta t}$$

$$I_{av} = C\frac{\Delta E}{\Delta t} \tag{1-28}$$

Rearranging Eq. 1-28,

$$\Delta E = \frac{I_{av}\,\Delta t}{C} \tag{1-29}$$

For one complete cycle of input voltage

$$\Delta t = \frac{1}{f} \tag{1-30}$$

Combining Eqs. 1-29 and 1-30

$$\Delta E = \frac{I_{av}}{Cf} \tag{1-31}$$

and, if

$$I_{av} = \frac{E_{av}}{R}$$

then

$$\Delta E = \frac{E_{av}}{fRC} \tag{1-32}$$

Referring to Fig. 1-9 we can define E_{av} as

$$E_{av} = E_m - \frac{\Delta E}{2} \tag{1-33}$$

Now, combining Eq. 1-32 with Eq. 1-33,

$$E_{av} = E_m - \frac{E_{av}}{2fRC} \tag{1-34}$$

and, by rearrangement,

$$E_{av} = E_m \left(\frac{2fRC}{2fRC + 1}\right) \tag{1-35}$$

It is apparent from Eq. 1-34 that if the *RC* time constant of the load is large, the average voltage approaches E_m.

To obtain some idea of the maximum diode current assume that at ωt_1, the start of the conduction period, $e_L = E_m - \Delta E$. It can be stated under these conditions, that

$$\omega t_1 = \sin^{-1}\left(\frac{E_m - \Delta E}{E_m}\right) \tag{1-36}$$

From Eq. 1-21

$$i_D = \frac{E_m}{R} \sin \omega t_1 + \omega C E_m \cos \omega t_1 \tag{1-37}$$

By substituting the value of ωt_1 found in Eq. 1-36 into Eq. 1-37, it is possible to solve for i_D at the start of the conduction period. Reference to Fig. 1-7(c) reveals that the peak diode current occurs at this angle.

The effective value of the a-c ripple voltage may be computed after assuming that the ripple has the triangular waveshape shown in Fig. 1-10. Since the average value of an a-c voltage is zero, the zero reference for this

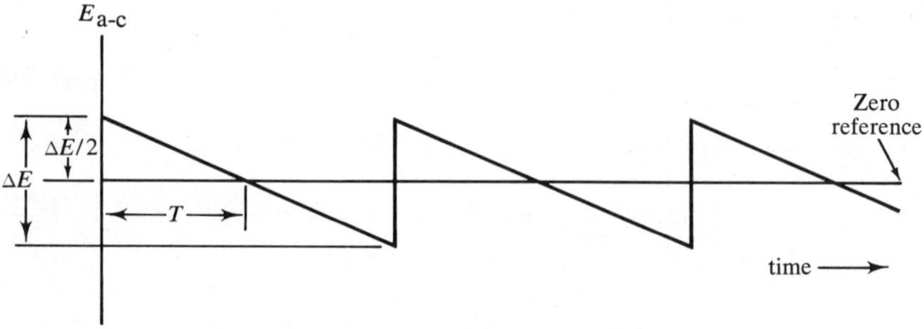

FIG. 1-10. Sawtooth approximation of a-c ripple in an *RC* load.

wave can be taken at $\Delta E/2$. This reference establishes an a-c component with an average value of zero. The instantaneous value of the wave then becomes

$$e = \frac{\Delta E}{2} - Mt \qquad 0 \leqq t \leqq T \tag{1-38}$$

where M = the slope of the voltage wave = $\Delta E/2T$. The effective value of this a-c voltage is derived as follows:

$$E_{a-c} = \sqrt{\frac{1}{T}\int_0^T \left(\frac{\Delta E}{2} - Mt\right)^2 dt}$$

$$E_{a-c} = \sqrt{\frac{1}{T}\int_0^T \left(\frac{\Delta E^2}{4} - \Delta EMt + M^2t^2\right) dt}$$

$$E_{a-c} = \sqrt{\frac{1}{T}\left[\frac{\Delta E^2 T}{4} - \frac{\Delta EMT^2}{2} + \frac{M^2T^3}{3}\right]_0^T}$$

$$E_{a-c} = \sqrt{\frac{\Delta E^2}{4} - \frac{\Delta EMT}{2} + \frac{M^2T^2}{3}} \tag{1-39}$$

Substituting $M = \Delta E/2T$ into Eq. 1-39,

$$E_{a-c} = \sqrt{\frac{\Delta E^2}{4} - \frac{\Delta ET}{2}\left(\frac{\Delta E}{2T}\right) + \frac{T^2}{3}\left(\frac{\Delta E^2}{4T^2}\right)}$$

This equation reduces to

$$E_{a-c} = \frac{\Delta E}{2\sqrt{3}} \tag{1-40}$$

which is the effective value of the a-c ripple voltage.

The equations just derived will next be used in solving a typical capacitor-filtering problem.

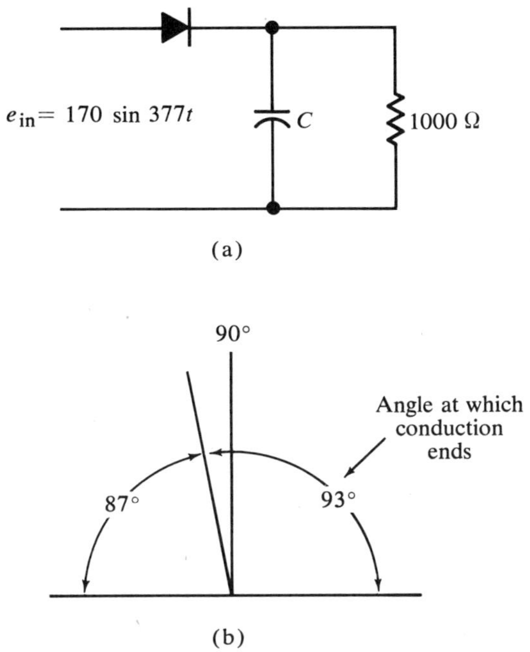

FIG. E1-2. (a) Half-wave rectifier circuit; (b) ωt_2 in the second quadrant.

Example 1-2 For the half-wave rectifier circuit of Fig. E1-2(a) a 10 percent ripple factor is desired. Calculate the:

1. value of the capacitor needed for a 10 percent ripple factor
2. average d-c voltage
3. effective value of the a-c ripple voltage
4. rms value of the combined d-c voltage and a-c ripple voltage
5. peak current through the diode
6. conduction time for the diode
7. average current through the diode
8. average power delivered to the load

1. Solve for the capacitor needed.
   From the definition of ripple and Eq. 1-16,

$$\frac{E_{\text{a-c}}}{E_{\text{av}}} = 0.1$$

$$E_{\text{a-c}} = 0.1E_{\text{av}}$$

Eq. 1-32 $$\Delta E = \frac{E_{av}}{fRC}$$

Eq. 1-40 $$E_{a\text{-}c} = \frac{\Delta E}{2\sqrt{3}}$$

Combining Eqs. 1-32 and 1-40,

$$2\sqrt{3}\,E_{a\text{-}c} = \frac{E_{av}}{fRC}$$

since $E_{a\text{-}c} = 0.1E_{av}$

$$2\sqrt{3}(0.1E_{av}) = \frac{E_{av}}{fRC}$$

Then,

$$C = \frac{1}{0.2\sqrt{3} \times 60 \times 1000}$$

$$C = 50\ \mu\text{F}$$

2. Calculate the average d-c voltage.
   Using Eq. 1-35

$$E_{av} = E_m\left(\frac{2fRC}{2fRC + 1}\right)$$

$$E_{av} = 170\left(\frac{2 \times 60 \times 1000 \times 50 \times 10^{-6}}{2 \times 60 \times 1000 \times 50 \times 10^{-6} + 1}\right)$$

$$E_{av} = 146\ \text{v}$$

3. Calculate the effective value of the a-c ripple.
   From Eq. 1-32

$$\Delta E = \frac{E_{av}}{fRC}$$

$$\Delta E = \frac{146 \times 10^6}{60 \times 1000 \times 50}$$

$$\Delta E = 49\ \text{v}$$

   From Eq. 1-40

$$E_{a\text{-}c} = \frac{\Delta E}{2\sqrt{3}}$$

$$E_{a\text{-}c} = \frac{49}{2\sqrt{3}}$$

$$E_{a\text{-}c} = 14\ \text{v}$$

Check: From the specified requirement of a 10 percent ripple factor,

$$E_{\text{a-c}} = 0.1E_{\text{av}}$$

$$E_{\text{a-c}} = 14.6 \text{ v}$$

the calculated factor of $E_{\text{a-c}}$ meets this requirement.

4. Calculate the rms value of the voltage.
From Eq. 1-17

$$E_{\text{rms}} = \sqrt{(E_{\text{av}})^2 + (E_{\text{a-c}})^2}$$

$$E_{\text{rms}} = \sqrt{(146)^2 + (14.1)^2}$$

$$E_{\text{rms}} = 147 \text{ v}$$

5. To calculate the peak current through the diode, Eq. 1-21 is used. The peak current occurs when the diode begins to conduct, that is, at ωt_1.

$$i_D = \frac{E_{\text{m}}}{R} \sin \omega t_1 + \omega C E_{\text{m}} \cos \omega t_1$$

From Eq. 1-36

$$\omega t_1 = \sin^{-1}\left(\frac{E_{\text{m}} - \Delta E}{E_{\text{m}}}\right)$$

$$\omega t_1 = \sin^{-1}\left(\frac{170 - 48.6}{170}\right)$$

$$\omega t_1 = 50°$$

Substituting this value of ωt_1 into Eq. 1-21,

$$i_D = \frac{E_{\text{m}}}{R} \sin 50° + \omega C E_{\text{m}} \cos 50°$$

$$i_D = \frac{170}{1000}(0.75) + (6.28)(60)(50 \times 10^{-6})(170)(0.64)$$

$$i_D = 2.1 \text{ A}$$

The maximum value of i_D is 2.1 A, which occurs at ωt_1.

6. Find the diode conduction time.
The diode conducts from ωt_1 to ωt_2 and ωt_1 has been found to be 50°. From Eq. 1-26

$$\omega t_2 = \tan^{-1}(-\omega RC)$$

$$\omega t_2 = \tan^{-1} - (6.28 \times 60 \times 50 \times 10^{-6} \times 1000)$$

$$\omega t_2 = \tan^{-1} - 18$$

Since the tangent is negative, the angle is in the second quadrant. To find the cut-off angle ωt_2, subtract the angle found from 180°. Thus, ωt_2, the angle where conduction ends = 180° − 87° = 93°.

The conduction angle ϕ is, therefore,

$$\phi = 93° - 50°$$

$$\phi = 43°$$

and the conduction time t_c becomes

$$t_c = 43° \frac{\frac{1}{60} \text{ sec}}{360°}$$

$$t_c = 0.002 \text{ sec}$$

7. Calculate the average diode current during conduction.

   While the diode is conducting it must supply the load current to the resistor and also the current required to charge the capacitor. The charge accumulated on the capacitor during the conduction portion of each cycle is equal to the average capacitor current during discharge multiplied by the discharge period. Since $E_{av} = 146$ v,

$$I_{av} = \frac{E_{av}}{R} = 0.146 \text{ A}$$

   and

$$Q = 0.146 \text{ A}(\tfrac{1}{60} \text{ sec}) = 0.00245\ C$$

   Since the diode must supply this charge in 0.002 sec,

$$I_{D_{av}} = j\frac{0.00245\ C}{0.002 \text{ sec}} + \frac{E_{av}}{R}$$

$$I_{D_{av}} = j\ 1.2 \text{ A} + 0.146 \text{ A}$$

   then

$$I_{D_{av}} = 1.21 \text{ A}$$

   which is the average diode current during the conduction period.

8. The power delivered to the load is

$$P_{av} = \frac{(E_{rms})^2}{R}$$

   or

$$P_{av} = \frac{(147)^2}{1000} = 21.6 \text{ W}$$

1-7 THE HALF-WAVE RECTIFIER WITH AN *RL* LOAD

When an inductive load is connected to a rectifier circuit, the inductor stores energy in its magnetic field during the time current is increasing and releases this energy while the current is decreasing. Since any practical inductor has some resistance associated with it, we will omit the case of a pure inductor and deal only with the practical case of an inductive and resistive (*RL*) load.

Figure 1-11(b) shows the waveforms for current and voltage in the *RL* circuit. During the time from 0 to $\pi/2$, the input voltage is positive in direction, and the induced voltage in the coil tends to oppose any increase in current through the load. When the input voltage passes its maximum value at $\pi/2$, the polarity of the induced voltage reverses and opposes any change in current.

The magnitude of the load current increases from $\pi/2$ to π, as shown. After the input voltage becomes negative, current continues in the forward direction through the rectifier until the negative value of e_{in} exceeds the positive value of e_L. At that point, the voltage across the rectifier is reversed, and no current exists. The conduction angle ϕ, for the period of negative input voltage, depends upon the values of L and R and approaches 180° in a pure inductive circuit. Conduction can thus be extended over a full 360°. When current flow is maintained, however, the average load voltage is reduced, as shown in the following analysis.

The average d-c voltage is

$$E_{\text{av}} = \frac{E_{\text{m}}}{2\pi}\int_0^{\pi+\phi} \sin \omega t(d\omega t)$$

where $\pi + \phi$ equals the total conduction angle of current through the diode. By integration,

$$E_{\text{av}} = \frac{E_{\text{m}}}{2\pi}\Big[-\cos \omega t\Big]_0^{\pi+\phi}$$

which reduces to

$$E_{\text{av}} = \frac{E_{\text{m}}}{2\pi}[1 - \cos(\pi + \phi)] \tag{1-41}$$

It can be seen from Eq. 1-41 that as ϕ approaches 180°, E_{av} approaches zero. This reduction of average load voltage limits the use of simple inductive filtering.

The equation for the instantaneous value of current in the circuit may be written as

$$i_D = \frac{E_{\text{m}}}{\sqrt{R^2 + \omega^2 L^2}}[\sin(\omega t - \Psi) + e^{(-Rt/L)} \sin \Psi]$$

where

$$\Psi = \tan^{-1}\left(\frac{\omega L}{R}\right) \tag{1-42}$$

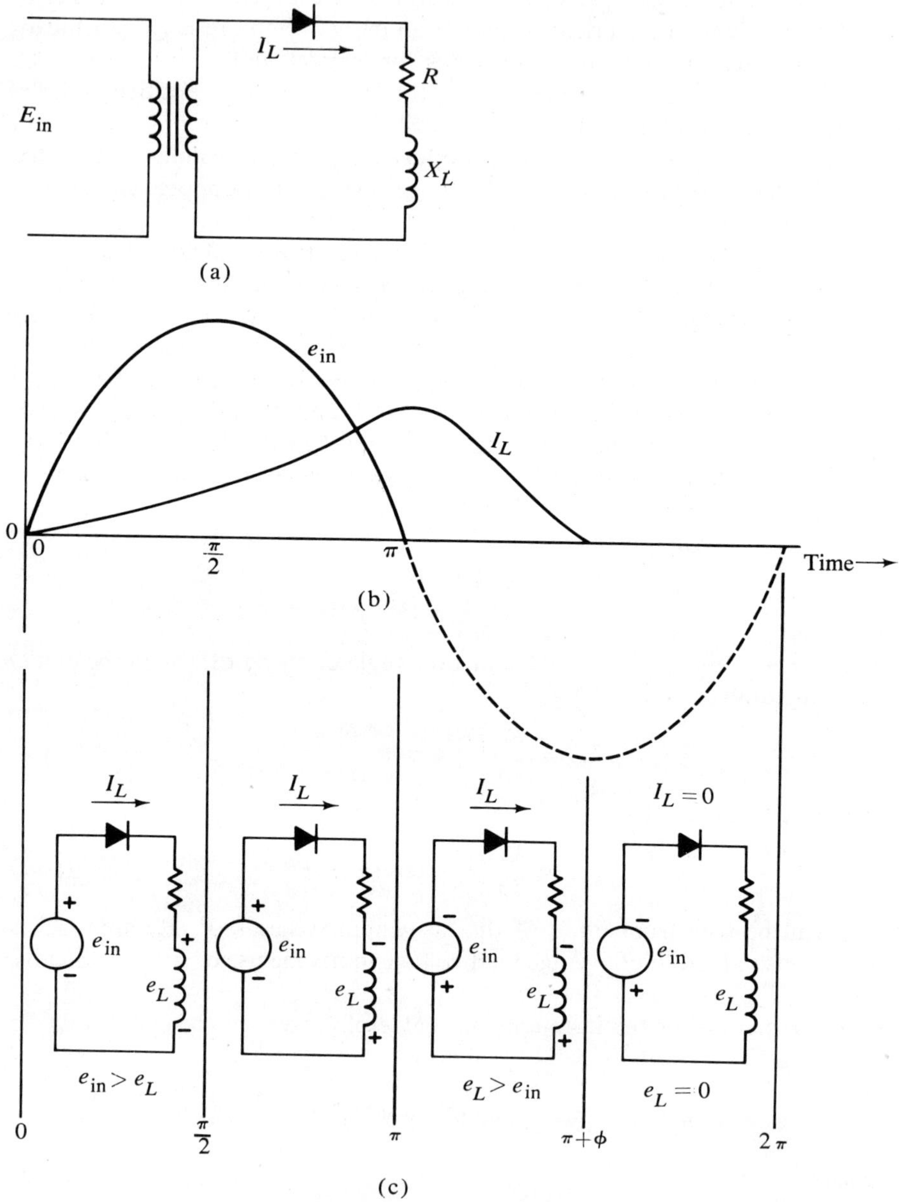

FIG. 1-11. (a) Half-wave rectifier with *RL* load; (b) Current- and voltage-waveforms; (c) Instantaneous relationships of *I* and *e*.

The conduction angle $(\pi + \phi)$ depends upon $\omega L/R$ and, from Eq. 1-41,

$$I_{av} = \frac{E_m}{2\pi R}[1 - \cos(\pi + \phi)] \tag{1-43}$$

Eq. 1-43 shows that the average value of current decreases as the value of inductance is increased.

Example 1-3 For the half-wave rectifier with an *RL* load shown below, compute the average load voltage (a) with the series inductance and (b) without the series inductance. $\phi = 50°$ when the inductance is added to the circuit.

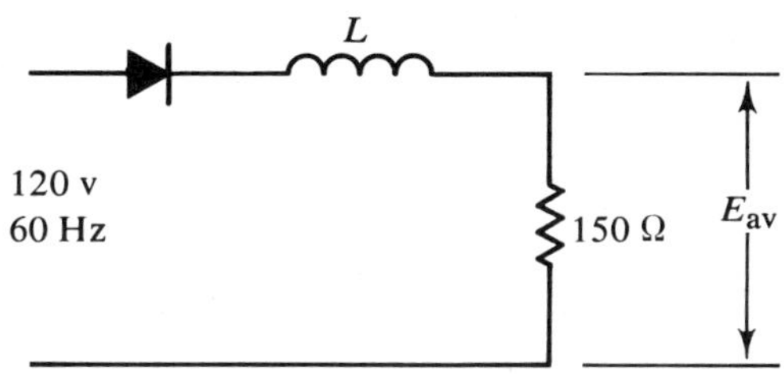

FIG. E1-3. Half-wave rectifier with *RL* load.

The average voltage without use of a choke may be calculated from Eq. 1-5.

$$E_{av} = \frac{E_m}{\pi} = 54 \text{ v}$$

Eq. 1-41 is used to compute the average voltage when a choke is used.

$$E_{av} = \frac{E_m}{2\pi}[1 - \cos(\pi + \phi)]$$

$$E_{av} = \frac{E_m}{2\pi}[1 - \cos(230°)]$$

$$E_{av} = \frac{E_m}{2\pi}(1.63)$$

$$E_{av} = 44 \text{ v}$$

The use of a choke has reduced the average voltage.

1-8 THE FULL-WAVE RECTIFIER WITH A RESISTIVE LOAD

The full-wave rectifier circuit is more widely used than the half-wave rectifier and is shown in Fig. 1-12. The full-wave circuit has the advantage of supplying

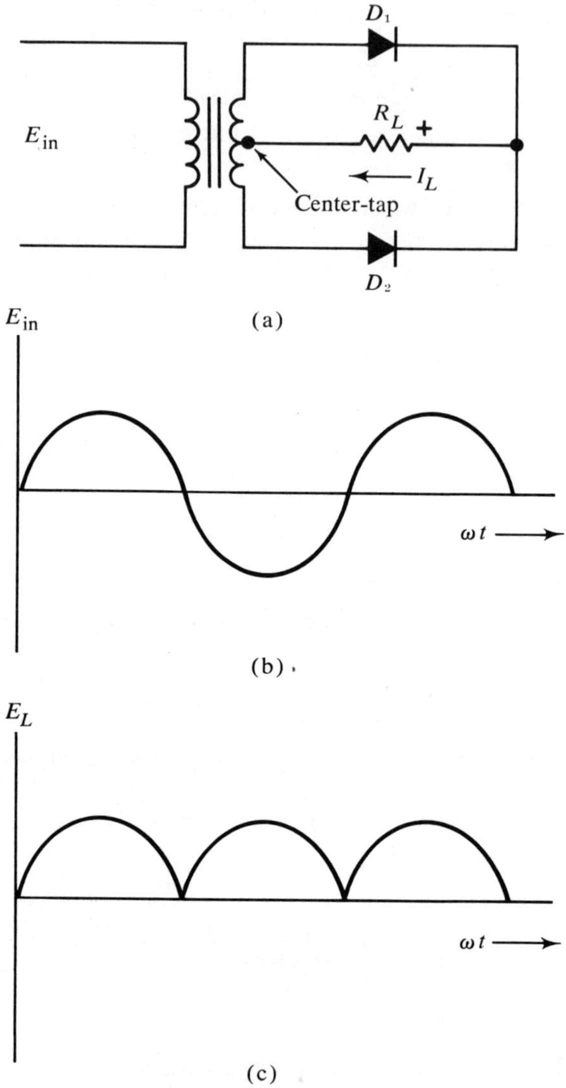

FIG. 1-12. (a) Full-wave rectifier circuit using center-tapped transformer; (b) Input voltage; (c) Load voltage.

current to the load during the entire cycle of the incoming wave. Therefore, a higher average current with a lower ripple factor is supplied by the full-wave circuit than by the half-wave rectifier. When the upper end of the transformer secondary is positive with respect to the center tap, the direction of current is through diode 1 in the forward direction, through the load resistor R_L, into the transformer center tap. Since no current passes through diode 2 in the reverse direction, the total rectified current passes through the load. When the lower end of the secondary is positive, current passes through diode 2 in the forward direction, through load resistor R_L, and back to the transformer center-tap. Thus, the direction of current through the load is always the same, that is, toward the transformer center tap, producing a load current having the waveform shown in Fig. 1-12(c). The average d-c voltage of this circuit is found by direct integration as follows:

$$E_{av} = \frac{1}{\pi}\int_0^{\pi} E_m \sin \omega t(d\omega t)$$

which reduces to

$$E_{av} = \frac{2E_m}{\pi} \quad \textbf{(1-44)}$$

and is twice the value of the average voltage for a half-wave rectifier.

The peak current through each diode is the same as that through a half-wave rectifier, or

$$I_m = \frac{E_m}{R}$$

The rms voltage is also found by integration.

$$E_{rms} = \sqrt{\frac{1}{2\pi}\int_0^{2\pi} (E_m \sin \omega t)^2 d\omega t}$$

which reduces to

$$E_{rms} = \frac{E_m}{\sqrt{2}} \quad \textbf{(1-45)}$$

This value of E_{rms} is identical to the rms value of a sinusoidal voltage.

Average load power in the full-wave circuit can be expressed as

$$P_{av} = \left(\frac{E_m}{\sqrt{2}}\right)^2 \times \frac{1}{R}$$

or

$$P_{av} = \frac{(E_m)^2}{2R} \quad \textbf{(1-46)}$$

The full-wave rectifier with a center-tapped transformer requires a secondary transformer voltage that is twice the required load voltage. However, the average current in each leg of the transformer winding is one-half the average load current, permitting the use of the same size of wire as that required for a half-wave rectifier. If the Kirchhoff voltage equation is written around the loop of Fig. 1-12, it can be seen that the peak inverse voltage across each diode is twice the value of the peak load voltage. Only one-half of the secondary voltage appears across the load at any time, but the full secondary voltage is across each diode for one-half cycle.

The ripple factor of a full-wave rectifier is much less than that of a half-wave rectifier as noted by a visual comparison of the waveforms of each circuit. Ripple factor was defined previously as the ratio of the rms value of the a-c component of voltage to the average value of the d-c component. Since the full-wave rectifier has an average d-c current that is higher than that of a half-wave rectifier and the a-c component is smaller, the ripple factor is substantially less for a full-wave rectifier. See Table 1-1.

TABLE 1-1. Comparison of Ripple Factor for Half-Wave and Full-Wave Rectifiers

Ripple Factor for Half-Wave Rectifiers	Ripple Factor for Full-Wave Rectifiers
$E_{av} = \dfrac{E_m}{\pi}$	$E_{av} = \dfrac{2E_m}{\pi}$
$E_{rms} = \dfrac{E_m}{2}$	$E_{rms} = \dfrac{E_m}{\sqrt{2}}$
$E_{a\text{-}c} = \sqrt{\left(\dfrac{E_m}{2}\right)^2 - \left(\dfrac{E_m}{\pi}\right)^2}$	$E_{a\text{-}c} = \sqrt{\left(\dfrac{E_m}{\sqrt{2}}\right)^2 - \left(\dfrac{2E_m}{\pi}\right)^2}$
$\gamma = \sqrt{\dfrac{\left(\dfrac{E_m}{2}\right)^2 - \left(\dfrac{E_m}{\pi}\right)^2}{\left(\dfrac{E_m}{\pi}\right)^2}}$	$\gamma = \sqrt{\dfrac{\left(\dfrac{E_m}{\sqrt{2}}\right)^2 - \left(\dfrac{2E_m}{\pi}\right)^2}{\left(\dfrac{2E_m}{\pi}\right)^2}}$
$\gamma = \sqrt{\dfrac{\dfrac{1}{4} - \dfrac{1}{\pi^2}}{1/\pi^2}}$	$\gamma = \sqrt{\dfrac{\dfrac{1}{2} - \dfrac{4}{\pi^2}}{4/\pi^2}}$
$\gamma = \sqrt{\dfrac{\pi^2}{4} - 1}$	$\gamma = \sqrt{\dfrac{\pi^2}{8} - 1}$
$\gamma = \sqrt{1.48}$	$\gamma = \sqrt{0.23}$
$\gamma = 121$ percent	$\gamma = 48.7$ percent

Example 1-4 For the full-wave center-tapped bridge in Fig. 1-12 assume the full secondary voltage to be 250 v and R_L to be 1000 ohms. Calculate the:

1. average value of the output voltage
2. rms value of the output voltage
3. ripple factor
4. maximum current per diode
5. average current per diode

1. From Eq. 1-44:

$$E_{av} = \frac{2E_m}{\pi}$$

The maximum voltage across the load is one-half the maximum secondary voltage, that is,

$$E_{av} = \frac{2}{\pi}\left(\frac{250\sqrt{2}}{2}\right)$$

$$E_{av} = 112 \text{ v}$$

2. From Eq. 1-45:

$$E_{rms} = \frac{E_m}{\sqrt{2}}$$

$$E_{rms} = 125 \text{ v}$$

3. To find the ripple factor we must first compute $E_{a\text{-}c}$.

$$E_{a\text{-}c} = \sqrt{(E_{rms})^2 - (E_{av})^2}$$

$$E_{a\text{-}c} = \sqrt{(125)^2 - (112)^2}$$

$$E_{a\text{-}c} = 55.8 \text{ v}$$

$$\gamma = \frac{55.8 \text{ v}}{112 \text{ v}} = 0.487$$

4. The maximum current per diode is:

$$I_m = \frac{E_m}{R}$$

$$I_m = \frac{176}{1000}$$

$$I_m = 0.176 \text{ A}$$

5. Since each diode conducts for only one-half cycle, the average current per diode is equal to one-half the average load current. Therefore,

$$\frac{I_{av}}{\text{diode}} = \frac{1}{2}\left(\frac{E_{av}}{R}\right)$$

$$\frac{I_{av}}{\text{diode}} = \frac{1}{2}\left(\frac{112 \text{ v}}{1000}\right) = 0.056 \text{ A}$$

1-9 THE FULL-WAVE RECTIFIER WITH AN *RC* LOAD

When a capacitor is connected in parallel with the load of a full-wave rectifier, as shown in Fig. 1-13(a), the average load current is increased and the percent of ripple is decreased. The analysis of this circuit is similar to the analysis of the half-wave rectifier previously discussed.

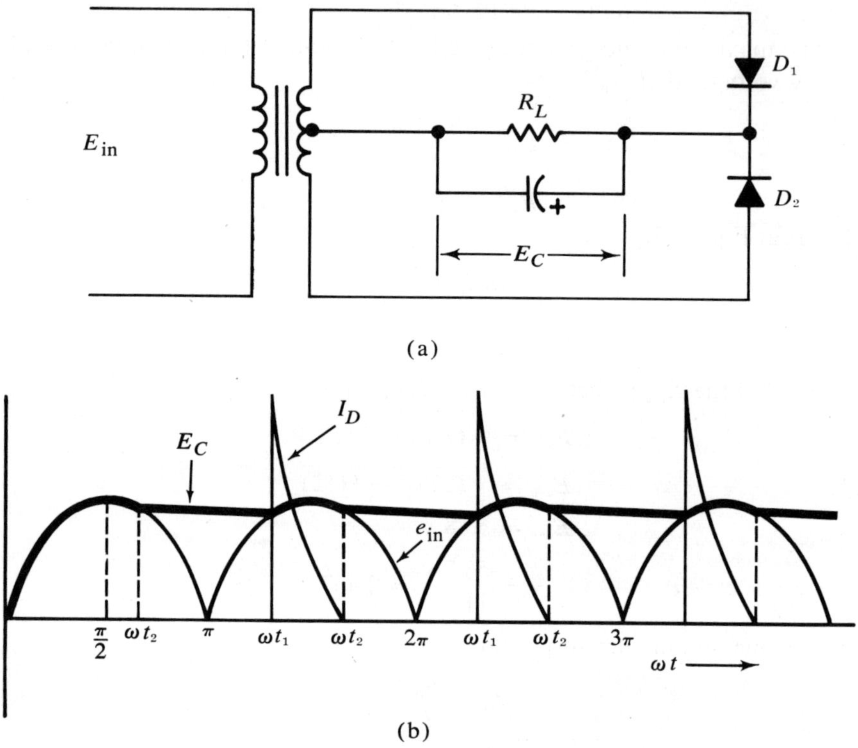

FIG. 1-13. (a) Full-wave rectifier with *RC* load; (b) Waveforms.

Assuming initially that the voltage at the upper end of the transformer secondary is going through zero in a positive direction, diode 1 conducts and the instantaneous voltage across the load and the capacitor is equal to e_{in} from 0 to $\pi/2$, as shown by the heavy lines in Fig. 1-13(b).

At $\pi/2$ the capacitor is charged to the value of the peak input voltage. As the input voltage falls after $\pi/2$ degrees, the capacitor voltage also falls. However, if the *RC* time constant of the load is large, the input voltage falls at a faster rate than the voltage on the capacitor, and at ωt_2 the capacitor voltage exceeds the input voltage. Diode 1 is now reverse biased and conduction ceases. Supplied by the charge stored in the capacitor, the load current continues until the voltage at the lower end of the secondary reaches the value at ωt_1. At ωt_1, the input voltage is greater than the capacitor voltage and diode 2 conducts. During their conduction periods, each diode must supply not only the load current but also the capacitor-charging current. As the capacitance of the load is increased, the peak current conducted by each diode increases but the conduction time decreases. The waveform of the current through each diode is shown in Fig. 1-13(b).

The average current from the source must equal the average load current; since the source supplies current for a short time only, the peak diode current must be greater than the peak load current. For a circuit containing a larger value of capacitor this peak current may be many times greater than the load current. (See Example 1-2.)

Since the capacitor is charged to the peak input voltage twice per cycle, the resultant average voltage across the load is higher than if a half-wave rectifier were used with the same *RC* load. To calculate the average voltage across the *RC* load, assume the capacitor charges instantly to the peak line voltage and then discharges at a constant rate for the next one-half cycle. After assuming that the capacitor charges instantly and discharges linearly, it can be said that the capacitor supplies the load current at all times, and that this current is constant because of the linear rate of discharge.

Using the method of analysis described for the half-wave rectifier,

$$I_{av} = C\frac{\Delta E}{\Delta t}$$

Because the capacitor charges twice per cycle

$$\Delta t = \frac{1}{2f} \qquad \textbf{(1-47)}$$

and

$$\Delta E = \frac{E_{av}}{2fRC} \qquad \textbf{(1-48)}$$

From Eq. 1-33,

$$E_{av} = E_m - \frac{\Delta E}{2}$$

or

$$E_{av} = E_m - \frac{E_{av}}{4fRC}$$

Rearranging,

$$E_{av} = E_m \left(\frac{4fRC}{4fRC + 1} \right) \tag{1-49}$$

Equation 1-49 shows that the average d-c voltage is a function of the RC time constant of the load and the source frequency.

A comparison of Eqs. 1-49 and 1-35 for the half-wave rectifier reveals that the average voltage will be higher in a full-wave rectifier with the same RC load. From Eqs. 1-48 and 1-32 it is obvious that ΔE and, therefore, $E_{a\text{-}c}$ are half as large in respective value for a full-wave rectifier as for a half-wave rectifier with a similar load.

Example 1-5 For the full-wave rectifier shown, calculate the:

1. average d-c voltage
2. ripple factor
3. power delivered to the load and compared with the half-wave rectifier of Example 1-2

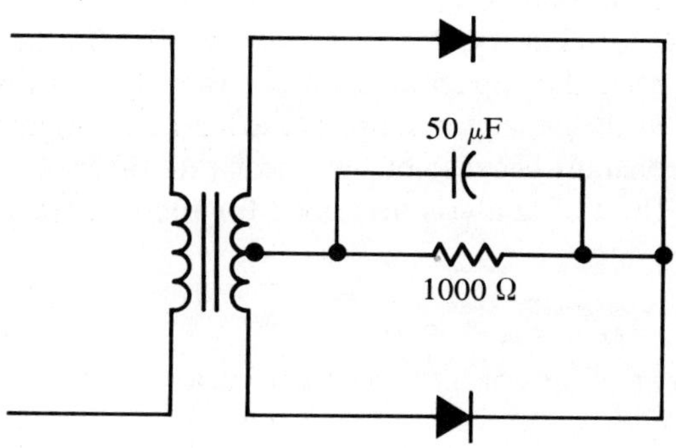

FIG. E1-5. Full-wave rectifier with load identical to that indicated in Problem 2.

1. Calculate E_{av}.
   From Eq. 1-49,

$$E_{av} = E_m \left(\frac{4fRC}{4fRC + 1} \right)$$

$$E_{av} = 170 \left(\frac{4 \times 60 \times 1000 \times 50 \times 10^{-6}}{4 \times 60 \times 1000 \times 50 \times 10^{-6} + 1} \right)$$

$$E_{av} = 158 \text{ v}$$

2. Calculate the ripple factor.

$$\gamma = \frac{E_{\text{a-c}}}{E_{\text{av}}}$$

From Eq. 1-48,

$$\Delta E = \frac{E_{\text{av}}}{2fRC}$$

$$\Delta E = \frac{158 \times 10^6}{2 \times 60 \times 1000 \times 50} = 26.4 \text{ v}$$

From Eq. 1-40,

$$E_{\text{a-c}} = \frac{\Delta E}{2\sqrt{3}}$$

$$E_{\text{a-c}} = \frac{26.4}{2\sqrt{3}} = 7.63 \text{ v}$$

$$\gamma = \frac{7.63}{158} \times 100 = 4.8 \text{ percent}$$

3. Calculate the power delivered to the load.

$$E_{\text{rms}} = \sqrt{(E_{\text{av}})^2 + (E_{\text{a-c}})^2}$$

$$E_{\text{rms}} = \sqrt{(158)^2 + (7.63)^2}$$

$$E_{\text{rms}} = 159 \text{ v}$$

$$P_{\text{av}} = \frac{(E_{\text{rms}})^2}{R}$$

$$P_{\text{av}} = \frac{(159)^2}{1000}$$

$$P_{\text{av}} = 25.2 \text{ W}$$

It will be helpful to the student to calculate the peak current through the diode, the conduction period for the diode, and the average diode current. The computation method is described in Example 1-2.

1-10 THE FULL-WAVE RECTIFIER WITH AN *RL* LOAD

It was shown in Sec. 1-7 that the current in a half-wave rectifier circuit with an inductive load continues for more than one-half cycle with an accompanying reduction of the average d-c voltage. It is now possible to analyze the full-wave rectifier with an RL load shown in Fig. 1-14.

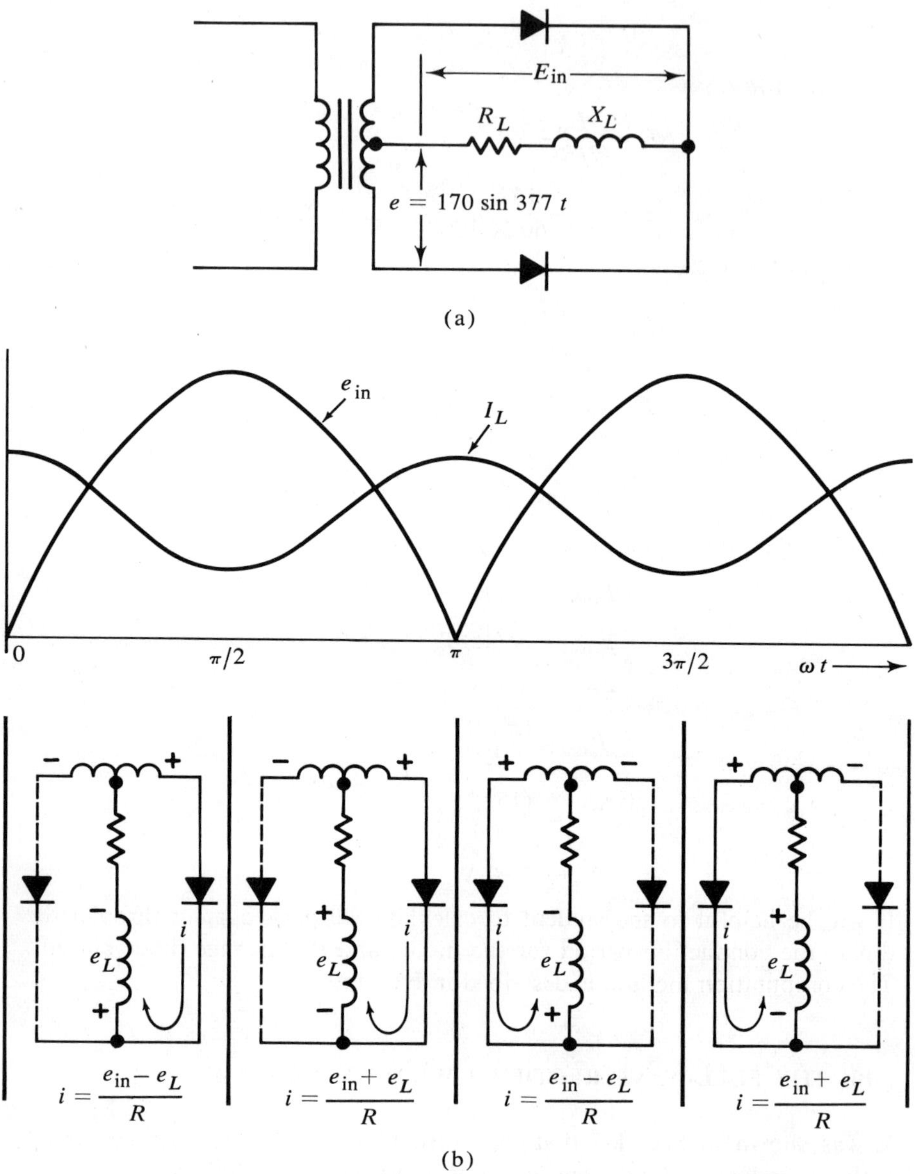

FIG. 1-14. (a) Full-wave rectifier with *RL* load; (b) Current- and voltage-waveforms.

When the supply voltage passes in a positive direction through zero, the voltage induced in X_L opposes the current, and energy is stored in the magnetic field of the inductor. From zero to $\pi/2$ the voltage across the load resistor is equal to the input voltage minus the voltage induced in the coil. Therefore, the current is

$$i = \frac{e_{\text{in}} - e_{\text{induced}}}{R} \qquad 0 \leqq \phi \leqq \frac{\pi}{2}$$

When the input voltage passes its maximum-positive value and starts to decay, the induced voltage opposes any change in current. The current can, therefore, be expressed as

$$i = \frac{e_{\text{in}} + e_{\text{induced}}}{R} \qquad \frac{\pi}{2} \leqq \phi \leqq \pi$$

The current is increased during this quarter-cycle as a result of the action of both the source and induced voltages. Even when the input voltage reaches zero, an induced voltage is present in the inductor, due to the decaying magnetic field. This induced voltage maintains current to the load and through both diodes at the instant that $e_{\text{in}} = 0$. As the supply voltage increases in the opposite direction, from π to $3\pi/2$, the induced voltage again opposes the current until the source voltage passes its negative peak and begins to decrease. At that time the induced and source voltages are joined, and the current in the load increases, as in the previous half-cycle. The important fact here is that the energy stored in the inductor prevents the load current from dropping to zero.

The voltage across the *RL* load is a rectified sine wave, and the average voltage may be expressed as

$$E_{\text{av}} = \frac{2E_{\text{m}}}{\pi} \tag{1-50}$$

The average current is

$$I_{\text{av}} = \frac{2E_{\text{m}}}{R\pi} \tag{1-51}$$

It can be assumed that the current is nearly constant and

$$P_{\text{av}} = E_{\text{av}}I_{\text{av}} = \frac{4(E_{\text{m}})^2}{R\pi^2} \tag{1-52}$$

Figure 1-14 shows that the a-c component in the output, because it is a nonsymmetrical wave, contains only even harmonics. The Fourier series for the wave form may be written as

$$e_{RL} = \frac{2E_{\text{m}}}{\pi} + \frac{2}{3}\left(\frac{2}{\pi} E_{\text{m}} \cos 2\,\omega t\right) \cdots \tag{1-53}$$

where ω is the source frequency, $2E_m/\pi$ is the average d-c voltage, and $[(2/3)(2E_m/\pi)] \cos 2\,\omega t$ is the magnitude of the second-harmonic component. Note that the output has no fundamental component from the a-c source.

The a-c current in the circuit is $\dfrac{E_{a\text{-}c}}{Z_L}$, and if we neglect all harmonics beyond the second,

$$I_{a\text{-}c} = \frac{\frac{2}{3}\left(\frac{2}{\pi}\right) E_m \frac{1}{\sqrt{2}}}{\sqrt{R_L^2 + (2\omega L)^2}}$$

or

$$I_{a\text{-}c} = \frac{4E_m(0.707)}{3\pi\sqrt{R_L^2 + (2\omega L)^2}}$$

which reduces to

$$I_{a\text{-}c} = \frac{0.3\ E_m}{\sqrt{R_L^2 + (2\omega L)^2}} \tag{1-54}$$

We define the ripple factor as

$$\gamma = \frac{I_{a\text{-}c}}{I_{av}}$$

Using $I_{a\text{-}c}$ from Eq. 1-54 and I_{av} from Eq. 1-51,

$$\gamma = \frac{0.3\ E_m}{\sqrt{R_L^2 + (2\omega L)^2}} \times \frac{R_L\pi}{2E_m}$$

This reduces to

$$\gamma = \frac{0.47\ R_L}{\sqrt{R_L^2 + (2\omega L)^2}}$$

or,

$$\gamma^2 = \frac{(0.47)^2 R_L^2}{R_L^2 + (2\omega L)^2}$$

and finally,

$$\gamma = \sqrt{\frac{(0.47)^2}{1 + \left(\frac{2\omega L}{R_L}\right)^2}} \tag{1-55}$$

Example 1-6 Refer to Fig. 1-14. If $R_L = 1000$ ohms, what value of inductance is needed to produce a 10 percent ripple factor?
From Eq. 1-55,

$$\gamma = \frac{0.47}{\sqrt{1 + \left(\frac{2\omega L}{R_L}\right)^2}}$$

$$0.1 = \frac{0.47}{\sqrt{1 + \left(\frac{2\omega L}{R_L}\right)^2}}$$

Squaring both sides,

$$0.01 = \frac{(0.47)^2}{1 + \left(\frac{2\omega L}{R_L}\right)^2}$$

$$0.01 + 0.01\left(\frac{2\omega L}{R_L}\right)^2 = (0.47)^2$$

Rearranging,

$$\frac{2\omega L}{R_L} = \sqrt{\frac{(0.47)^2 - 0.01}{0.01}}$$

$$\frac{2\omega L}{R_L} = 4.58$$

$$2\omega L = 4580$$

Solving for L,

$$L = \frac{4580}{2(377)}$$

$$L = 6.1 \text{ H}$$

1-11 THE FULL-WAVE RECTIFIER BRIDGE

A third, common rectifier circuit which, without a transformer, can supply a full-wave rectified output is the full-wave bridge rectifier. The circuit shown in Fig. 1-15 reduces the peak-inverse voltage per diode from the PIV of a center-tapped bridge but requires four rectifier elements instead of two.

When terminal 1 of the supply is positive, diodes 2 and 3 conduct, the current passing into the load at point A. Current through diodes 1 and 4 at this time is zero because both are reverse-biased. When terminal 2 of the supply is positive, diodes 1 and 4 conduct, diodes 2 and 3 acting as open circuits. Current is still directed into the load at point A, as shown. The peak-inverse voltage across the nonconducting diodes is E_m of the supply. Both the average and effective currents in the load are the same as those previously determined for a full-wave, center-tapped rectifier.

$$E_{av} = \frac{2E_m}{\pi}$$

and

$$E_{rms} = \frac{E_m}{\sqrt{2}}$$

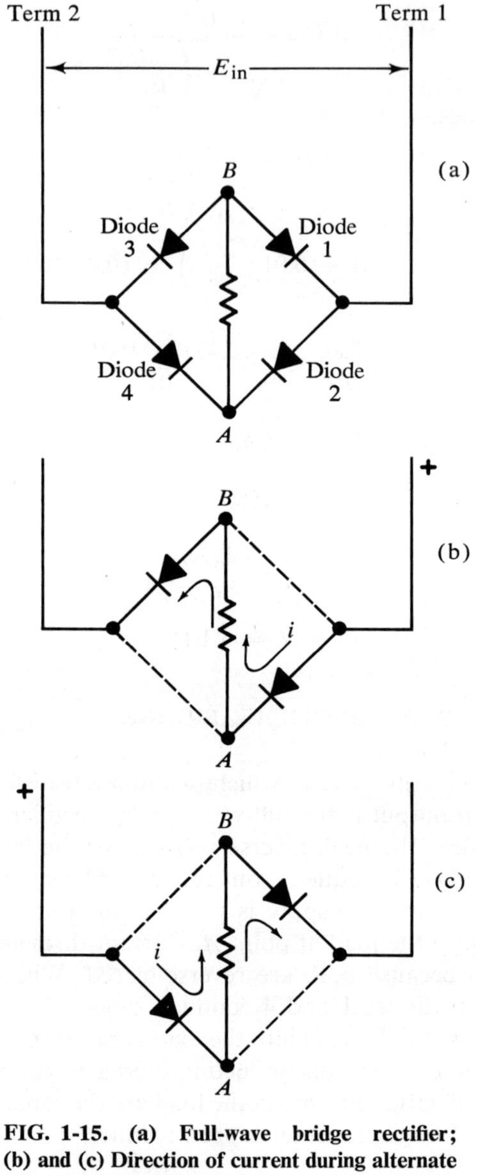

FIG. 1-15. (a) Full-wave bridge rectifier; (b) and (c) Direction of current during alternate half-cycles of input voltage.

The average and effective currents through each diode are one-half as great as in the load, because each diode conducts for only one-half cycle.

$$\frac{I_{av}}{\text{diode}} = \frac{I_m}{\pi} \tag{1-56}$$

and

$$\frac{I_{rms}}{\text{diode}} = \frac{I_m}{2\sqrt{2}} \tag{1-57}$$

The peak current in each diode is the same as that in the load.

$$I_m = \frac{E_m}{R}$$

This circuit has an advantage over a full-wave rectifier circuit with center-tapped transformer in that it requires only one-half the magnitude of supply voltage. Thus, the voltage rating of the transformer can be lower.

1-12 VOLTAGE-DOUBLERS AND VOLTAGE-TRIPLERS

When current demand is small, voltage-doublers and voltage-triplers may be used to supply voltages higher than the supply voltage. Many variations of doubler and tripler circuits are available, but all operate on the same principle. A capacitor is charged during one-quarter cycle to E_m of the input, and the voltage on the capacitor is added to the supply voltage in order to obtain 2 E_m. The process of adding the stored voltages on the capacitors can be continued until the desired voltage is obtained.

A typical voltage-doubler is shown in Fig. 1-16. When terminal B is positive, as shown in Fig. 1-16(a), diode 1 conducts, and capacitor C_1 charges to E_m. When terminal A is positive, the charge on capacitor C_1 is in series with the supply voltage and 2 E_m appears across capacitor C_2. Capacitor C_2 charges to 2 E_m with the polarity shown. The voltage across the load exhibits the waveform shown in Fig. 1-16(b). If the RC time constants R_LC_1 and R_LC_2 are large, the voltage on capacitor C_1 is maintained near E_m when the input is zero. As the value of the load resistor decreases, the average d-c voltage also decreases. The time constants R_LC_1 and R_LC_2 must, therefore, be large enough to maintain a charge for the period of a full cycle.

The same assumptions made regarding the half- and full-wave rectifiers may be used in predicting the average output voltage of a doubler, namely, that the capacitor charges instantaneously and discharges linearly. Then, referring to Fig. 1-16(c),

$$E_{av} = 2E_m - \frac{\Delta E}{2} \tag{1-58}$$

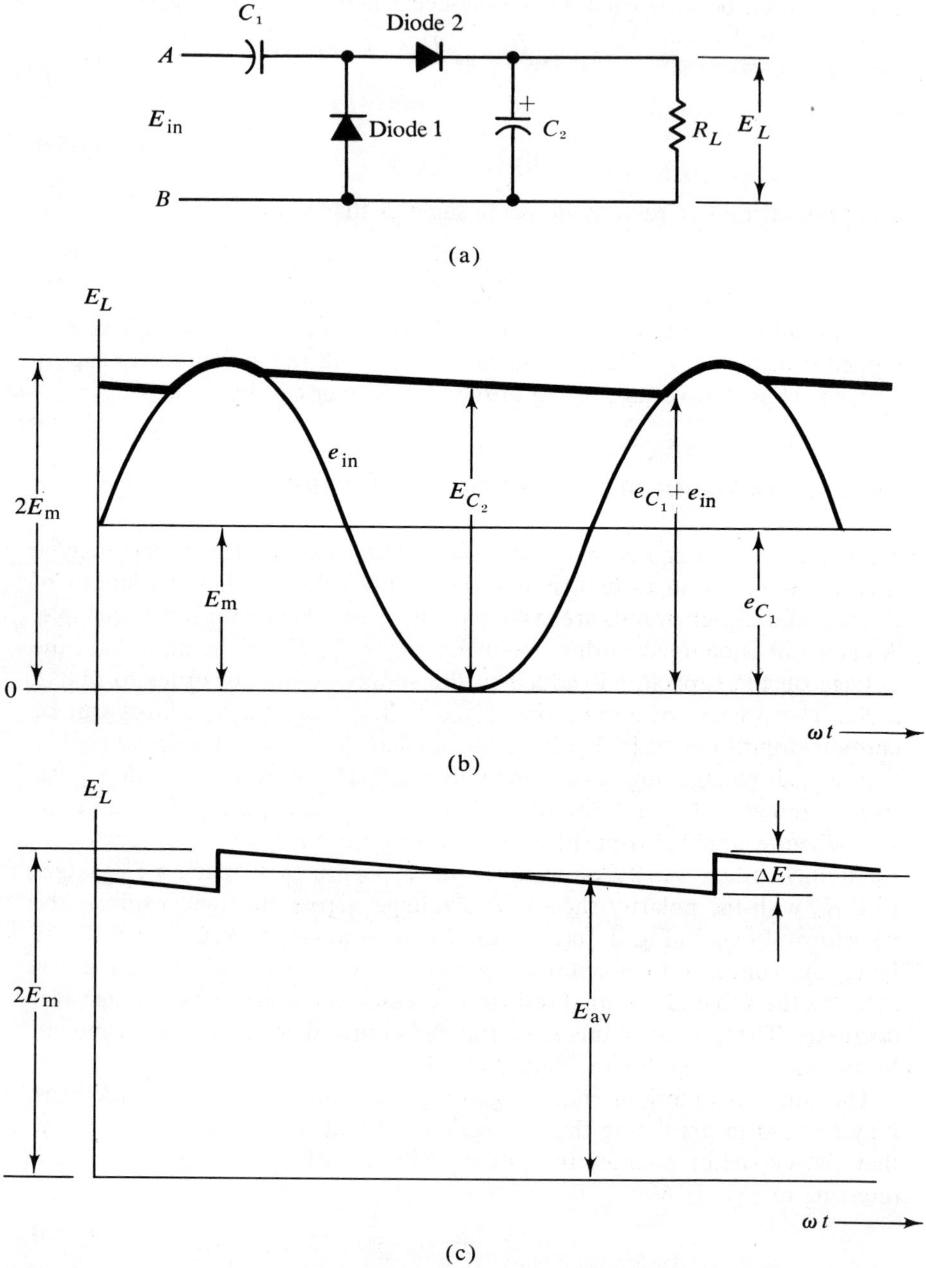

FIG. 1-16. (a) Voltage-doubler circuit; (b) Voltage waveforms; (c) A-C ripple voltage.

Using the same derivation technique as that used for the half-wave rectifier,

$$E_{av} = 2E_m \left(\frac{2fRC}{2fRC + 1} \right) \tag{1-59}$$

and

$$\Delta E = \frac{E_{av}}{fRC}$$

$$E_{a\text{-}c} = \frac{E_{av}}{2fRC\sqrt{3}}$$

The value of capacitor C_2 is used in the above equations as that capacitor is the one that supplies current to the load when diode 2 is not conducting.

Example 1-7 Calculate the capacitance needed in a voltage-doubler circuit to produce 200 v average across a 1,000-ohm resistor from a 120-v, 60-Hz source. Using Eq. 1-59

$$E_{av} = 2E_m \left(\frac{2fRC}{2fRC + 1} \right)$$

$$200 = 2(170) \left(\frac{2 \times 60 \times 1000 \times C}{2 \times 60 \times 1000 \times C + 1} \right)$$

Solving for C,

$$C = 12\ \mu\text{F}$$

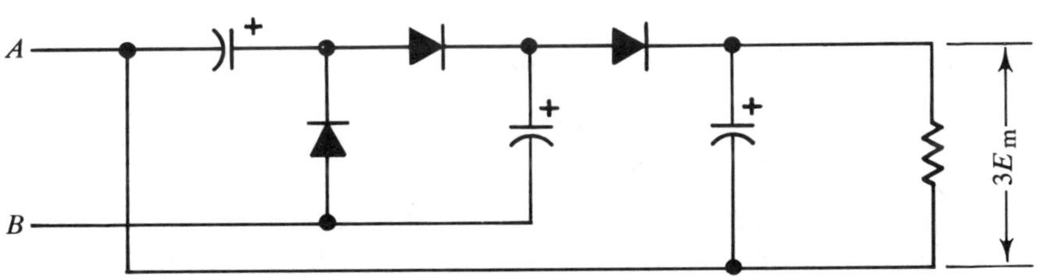

(a)

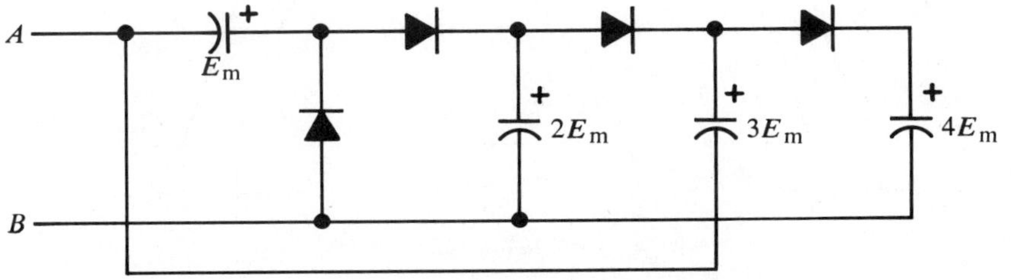

(b)

FIG. 1-17. (a) Voltage-tripler; (b) Voltage-quadrupler.

It is considered good design practice to make capacitor C_1 equal to 5 C_2 in the voltage-doubler circuit.

1-13 PRACTICAL-RECTIFIER CONSIDERATIONS

In the foregoing discussions it was assumed that a diode element conducts at exactly 0° and ceases conduction at 180°. In a practical rectifier, however,

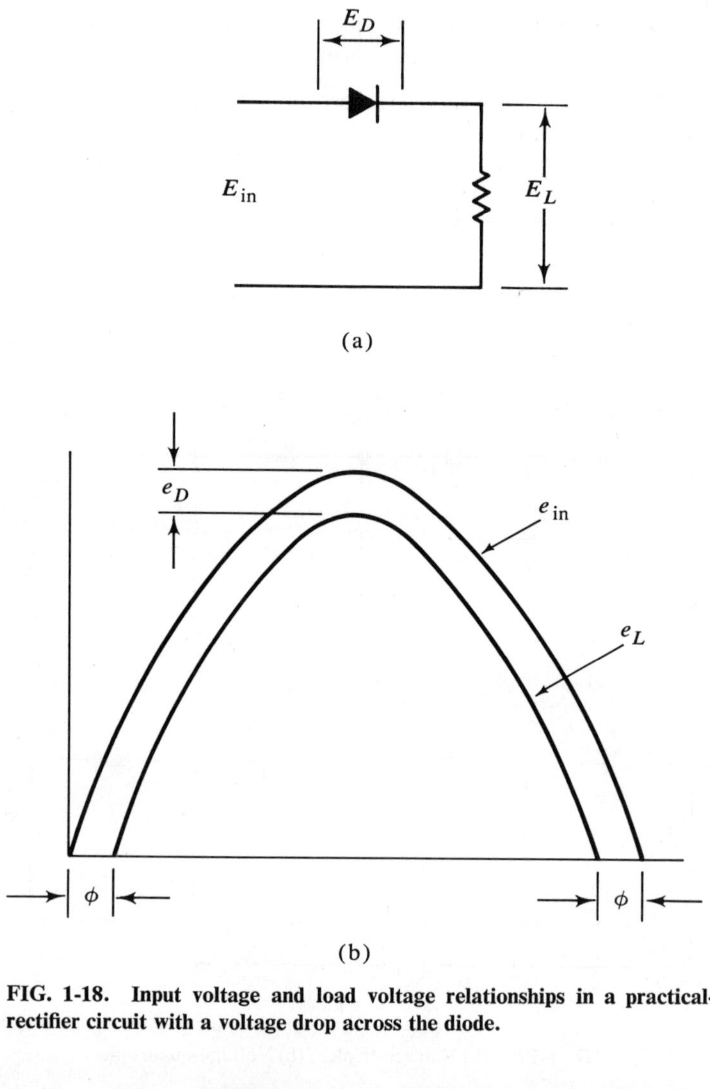

FIG. 1-18. Input voltage and load voltage relationships in a practical-rectifier circuit with a voltage drop across the diode.

conduction starts slightly beyond 0° and ceases before 180° are reached, as shown in Fig. 1-18. The angle ϕ is called the *cut-in* and *cut-out angle* and can be expressed as $\phi = \sin^{-1} E_D/E_m$ where E_D is the forward voltage drop across the diode. This voltage drop is larger in vacuum-tube rectifiers than in semiconductor diodes. The extent to which E_D must be considered when designing a rectifier depends upon the application. The forward-voltage drop causes a temperature rise in the diode element as a result of I^2R loss in the element. This loss can be expressed as

$$P = (E_D)^2/R_D \tag{1-60}$$

where E_D is the drop across the diode and R_D is the forward resistance of the diode. It can be seen from Eq. 1-60 that the greater the forward voltage drop across the diode, the greater is the power loss; and consequently, the greater is the temperature rise. Temperature rise in the rectifying element limits its current-carrying capacity and is the deciding factor in rating the device.

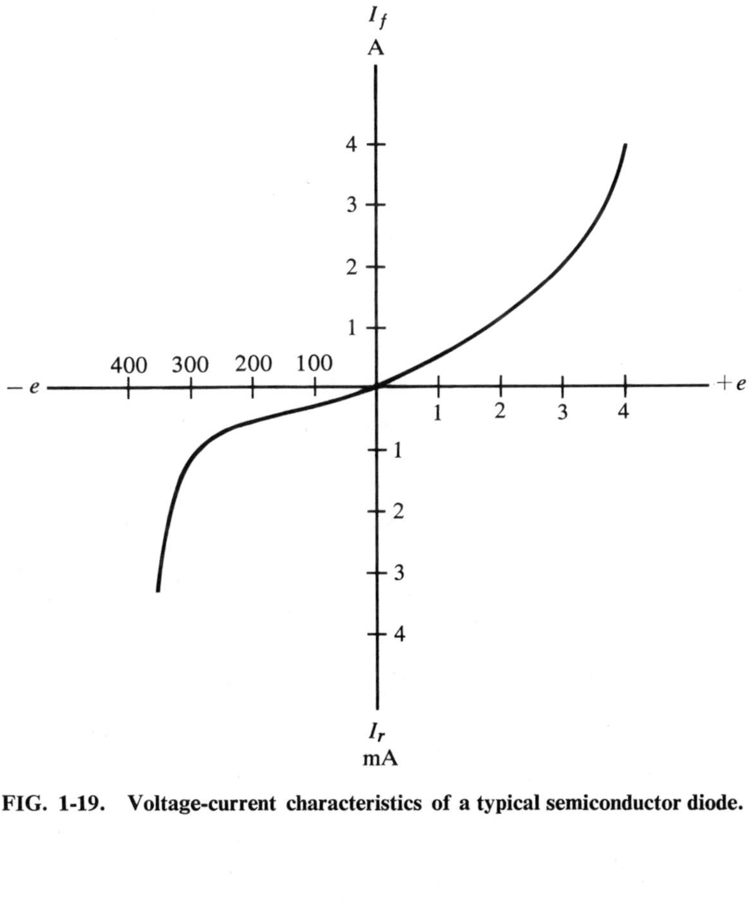

FIG. 1-19. Voltage-current characteristics of a typical semiconductor diode.

The percentage-voltage regulation of a rectifier is determined by the output voltage at no-load and full-load.

$$\text{Percent regulation} = \frac{\text{no-load voltage} - \text{full-load voltage}}{\text{full-load voltage}} \times 100$$

The percentage regulation is determined primarily by the output impedance of the rectifier circuit. The output impedance, in turn, is determined by the transformer regulation, voltage drop in the diodes, type of filter, etc. In general, as more filtering is inserted into a rectifier circuit, its percentage-regulation increases.

Figure 1-19 shows a typical curve for the forward and reverse characteristics of a semiconductor diode. From this curve it is possible to predict the power loss in the rectifying element for any particular value of current. By multiplying the average current through the diode by the voltage drop across the unit, we can obtain a good estimate of the expected heat rise.

Since the reverse-leakage current generally rises with temperature, lowering the efficiency of the unit, specified maximum temperatures for semiconductors should not be exceeded.

Example 1-8 Using the curve of Fig. 1-19, calculate the power loss, efficiency, and regulation of the half-wave rectifier shown.

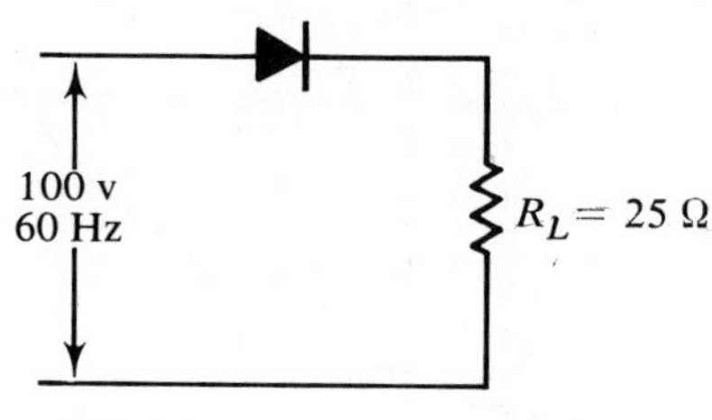

FIG. E1-8. Half-wave rectifier circuit.

$$E_{\text{av}} = \frac{100\sqrt{2}}{\pi} = 45 \text{ v}$$

and

$$I_{\text{av}} = \frac{45 \text{ v}}{25\ \Omega} = 1.8 \text{ A}$$

From this curve the voltage drop across the diode at 1.8 A = 3 v; the diode resistance is

$$3 \text{ v}/1.8 \text{ A} = 1.66\ \Omega$$

The power loss in the diode is calculated from Eq. 1-60.

$$P = (3)^2/1.66$$

$$P = 5.4 \text{ W}$$

The efficiency, eff, $= P_{\text{out}}/P_{\text{in}}$

$$\text{eff} = \frac{(1.8)^2 \times 25}{(1.8)^2 \times 25 + 5.4}$$

$$\text{eff} = \frac{81}{86.4} \times 100$$

$$\text{eff} = 93.8 \text{ percent}$$

$$\text{The percent regulation} = \frac{E_{NL} - E_{FL}}{E_{FL}} = \frac{45 - 42}{42} \times 100 = 6.7 \text{ percent}$$

1-14 FILTERS

A simple filter, such as a capacitor or an inductor in series with a load, gives a smoothing effect to the load current in a rectifier circuit.

Although the simple capacitor or inductor filter may be adequate for many applications, when a smoother output current is required, more elaborate filters must be used.

The filter shown in Fig. 1-20(a) consists of an inductor in series with the load and a capacitor across the load. It has been shown that the rectifier output voltage E_o consists of an average d-c voltage with an a-c component. The rectifier circuit is represented in Fig. 1-20(b). There, two sources of voltage are in series; the d-c source is represented by the dotted line in Fig. 1-20(c), and superimposed on this d-c voltage is the a-c wave shown by the solid line. The instantaneous value of output voltage varies from $E_{\text{d-c}} + E_{\text{p}}$ to $E_{\text{d-c}} - E_{\text{p}}$. The filter circuit should be able to reduce the a-c component of current in the load until the load current approaches a pure d-c value.

The frequency of the a-c component is equal to the line frequency for a half-wave rectifier and equal to twice the line frequency for a full-wave rectifier. The ripple factor of the power supply depends upon the peak value of the a-c component and is easier to reduce if this a-c voltage is at the higher frequency.

If the direct and alternating currents are supplied to the load by the rectifier, it can be seen that the inductor L offers little resistance to the direct current but provides opposition to the alternating current. The resistance to the direct current is simply the internal resistance of the coil, which can be very small. The resistance to the alternating current can be expressed as $2\,\pi fL$ where f is the frequency of the a-c component. If L is large, a large portion of the a-c component of the rectified voltage is dropped across it.

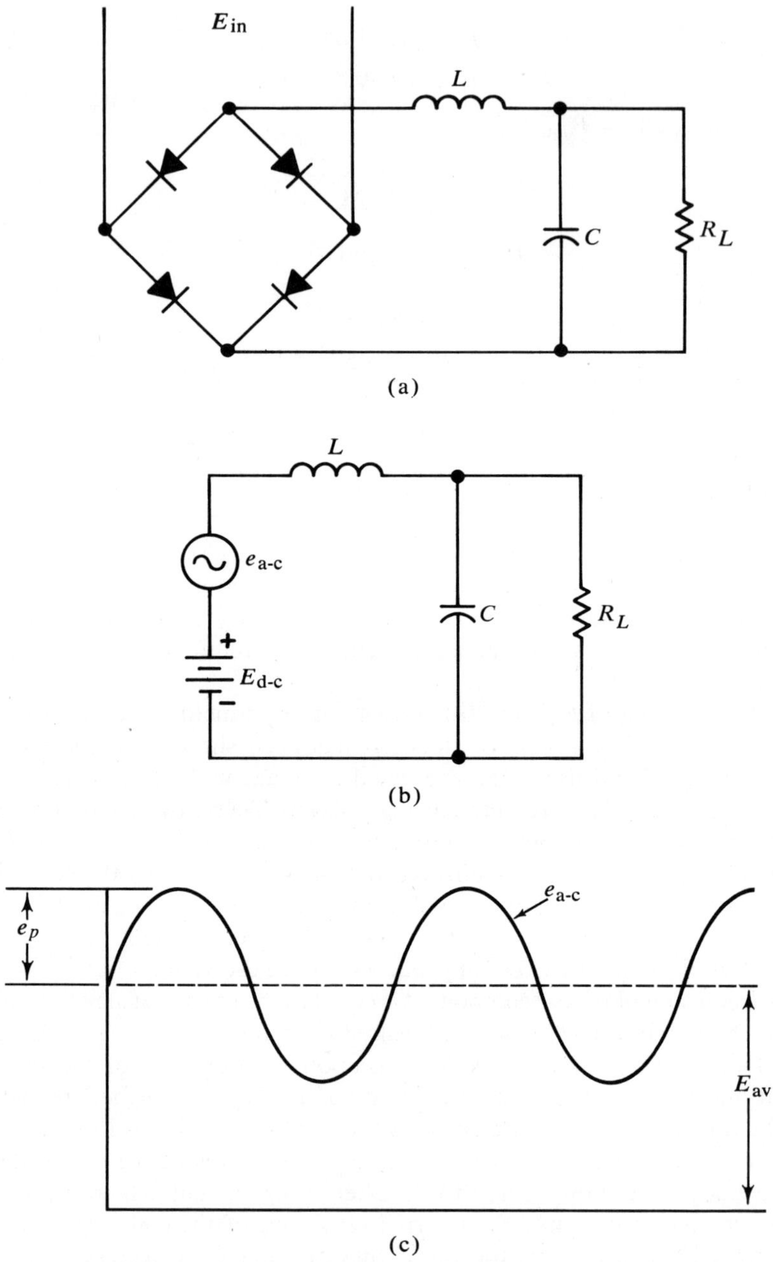

FIG. 1-20. (a) Rectifier circuit with choke-input filter; (b) Equivalent circuit for rectifier; (c) Voltage waveforms of the equivalent circuit.

Capacitor C_1 should be large enough to offer less opposition than the load resistor to the a-c component. If C is large, the alternating current meets little opposition in traveling back to the supply through C, resulting in reduced alternating current through the load.

This type of filter is referred to as a choke-input filter because the output of the rectifier is fed into an inductor which cuts off (chokes) the a-c components. Such a filter is used generally in applications requiring high power and a high ratio of average-to-peak power.

The filter acts to smooth the incoming voltage wave by storing energy in the magnetic field of the inductor at peak values of voltage and releasing this energy evenly to the load after the input voltage drops below the load voltage. The capacitor reduces the a-c component by charging to the peak voltage and discharging into the load resistor whenever the input voltage drops. This discharge of the capacitor tends to maintain direct current into the load.

The choke-input filter shown in Fig. 1-20(a) is used when the regulation of the power supply is of greater importance than the quantity of a-c ripple. When ripple in the output is the prime concern, the capacitor-input filter

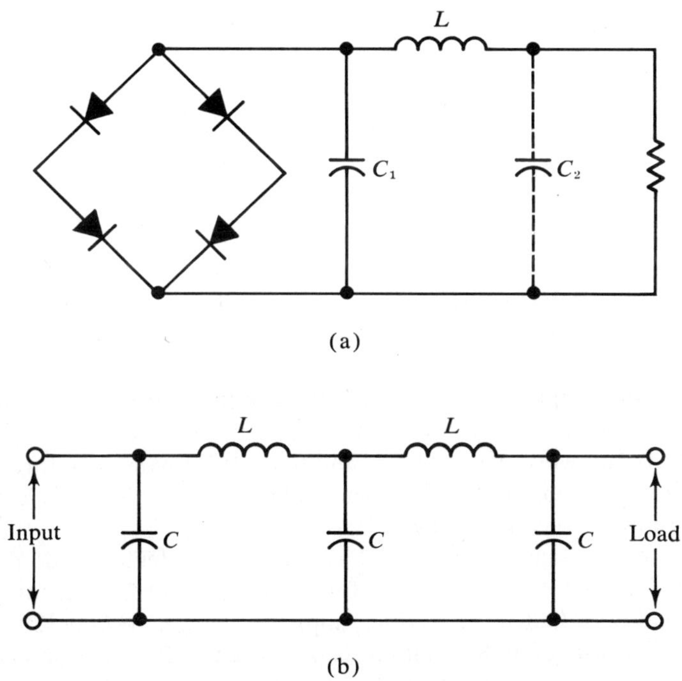

FIG. 1-21. Rectifier circuit with: (a) capacitive-input filter, pi (π) filter produced by adding capacitor C_2; (b) double-π filter.

shown in Fig. 1-21(a) is used. The capacitor-input filter reduces the amount of ripple for the same amounts of inductance and capacitance as in the filter illustrated in Fig. 1-20(a) but gives poorer regulation.

Because of cost factors, the *LC* filter may occasionally be replaced with a resistance input or *RC* filter. The resistance-input filter has a high *IR* drop and power loss and, therefore, is useful only in circuits with low current-drain. The capacitor in the *RC* filter circuit acts to smooth the incoming wave by charging to peak value and then discharging into the load when the supply voltage falls below the peak. To construct an *RC* filter it is necessary to replace the inductor *L* in Fig. 1-20(a) by a resistor of appropriate value.

The π filter is an *LC* filter with a capacitor added across the load terminals of the rectifier, as shown in Fig. 1-21(a). The added capacitor C_2 reduces the percentage of ripple in the output. The double filter shown in Fig. 1-21(b) gives still better filtering but with a reduction in regulation. The type of filter to be used is governed by the particular application.

Problems

1. Find the average values for the waveforms shown in Fig. P1-1, first by (a) graphic techniques and second by (b) integration techniques. In addition, calculate the rms values of the respective waveforms, and also find the ripple factor for each waveform.
2. A half-wave rectifier circuit has an a-c input voltage of 120 v. If a 200-ohm resistor is used for a load, calculate: I_m, I_{av}, I_{rms}, the ripple factor, and load power.
3. A rectifier circuit supplies an average current of 8 A to a load. If the ripple factor at the load is 30 percent, what is the rms value of the current?
4. Find the form factor for a waveform which has an rms value of 10 v and an average value of 8 v.
5. What is the frequency of the ripple voltage in the following rectifier circuits: (a) 400 Hz full-wave single-phase, (b) 60 Hz half-wave single-phase, (c) 60 Hz full-wave, three-phase, and (d) 400 Hz half-wave three-phase? Which rectifier circuit will have the highest ripple voltage if each is supplying power to equivalent *RC* loads?
6. The output voltage of a rectifier has a unity form factor. Find the ripple factor. Find the new ripple factor for an increase to 1.2 in the form factor.
7. The load on a half-wave rectifier, supplying power from a 120-v, 60-Hz source, consists of an 800-ohm resistor and a 50-μF capacitor. Calculate: (a) the angle at which the diode starts to conduct and (b) the angle at which the diode ceases conduction. Draw a vector diagram showing the angles. At what point does the current in the capacitor reverse?

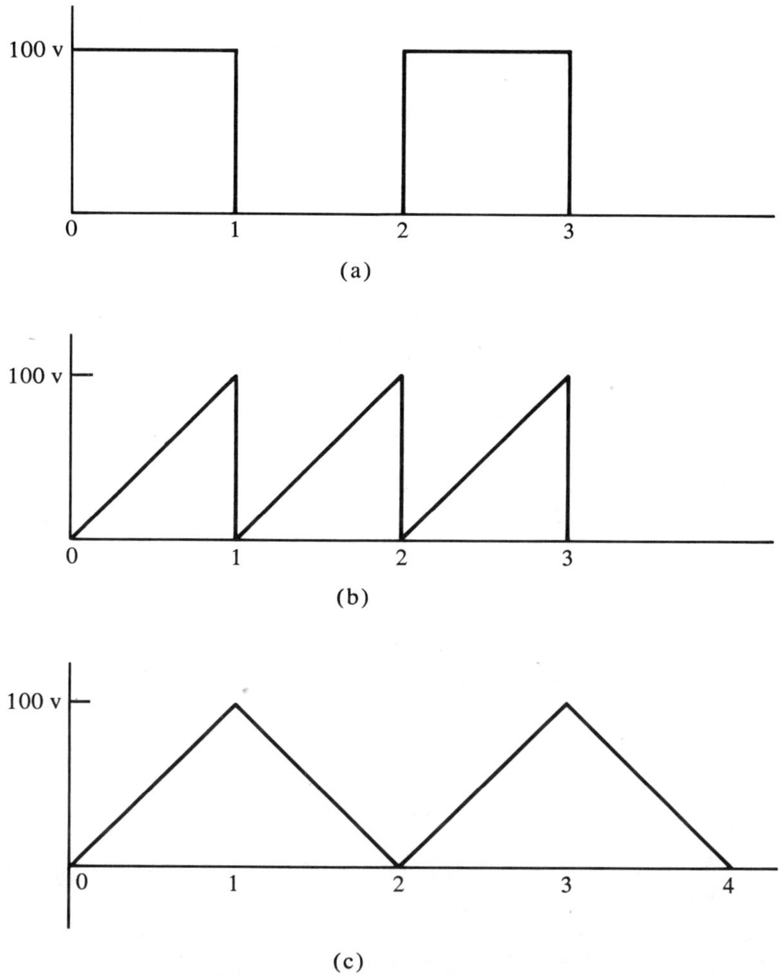

FIG. P1-1. Figure for Problem 1.

8. For the circuit of Problem 7 calculate the peak diode current.
9. What is the average current delivered to a load from a charged 100-μF capacitor if its voltage drops 30 v in 0.01 sec? Assume a linear discharge rate.
10. The average current in a 1000-ohm resistor connected to a half-wave rectifier is 0.2 A. If the ripple frequency is 800 Hz and the resistor is in parallel with a 10-μF capacitor, what is the magnitude of the voltage change across the capacitor?

11. What is the average voltage at the output of a full-wave rectifier fed from a 120-v, 60-Hz line, if the load consists of a 500-ohm resistor and a 30-μF capacitor in parallel?
12. Find the average voltage across the 2000-ohm load of a half-wave rectifier that is connected to a 240-v, 60-Hz source. A 10-μF capacitor is in parallel with the resistor.
13. For the circuit in Problem 12 find the angle during which the diode conducts, the peak current in the diode, and the average diode current during conduction.
14. The output voltage of a rectifier varies from a maximum of 240 v to a minimum of 180 v. Find the effective value of the a-c component of this rectifier. Find the average and effective values of the voltage and ripple factor of this rectifier.
15. A half-wave rectifier delivers an average voltage of 100 v to a load. The peak-input voltage is 130 v, and the source frequency is 400 Hz. If the load consists of a 2000-ohm resistor and a capacitor calculate: (a) the value of the capacitor, (b) the average and effective load currents, (c) the effective value of the a-c ripple voltage, (d) the number of degrees during which the diode conducts, (e) the peak current in the diode, and (f) the power delivered to the load.
16. Calculate the value of the capacitor needed to produce a 10-percent ripple factor on the output of a half-wave rectifier with a 2000-ohm load. The rectifier is fed from a 140-v, 60-Hz source.
17. A rectifier delivers 10 W to a load consisting of a 1,000-ohm resistor and a capacitor. The ripple factor of the voltage across the load is 20 percent. Find the average load voltage and the peak-to-peak value of the a-c ripple voltage.
18. A half-wave rectifier with an *RL* load is connected to a 120-v, 60-Hz source. If the diode conducts for 210°, what is the average voltage across the resistive load?
19. What power is delivered to a 1000-ohm resistor connected to a full-wave rectifier fed from a 120-v, 60-Hz line?
20. If the circuit of Prob. 19 were changed to a half-wave rectifier, how much power would be delivered to the load?
21. Calculate the ripple factor, average and rms values of voltage, and conduction time for the diodes if a 1000-ohm resistor and a 5-μF capacitor are connected to the output of a full-wave rectifier. The frequency of the 120 v input is 60 Hz. Repeat for a half-wave rectifier.
22. Calculate the ripple factor at the output of a full-wave, center-tapped, transformer rectifier circuit. The maximum voltage across the load is 240 v, and the load consists of a 200-ohm resistor in parallel with a 75-μF

capacitor. The rectifier is connected to a 60-Hz source. If the capacitor is decreased to 50 μF, calculate the new ripple factor. Repeat for 10 μF.

23. Referring to Fig. 1-14, what is the ripple factor in the output of the rectifier circuit if $R_L = 1000$ ohms and $L = 0.5$ Hz? Repeat for $L = 0.7$ Hz.
24. A full-wave rectifier delivers an average current of 100 A to an *RC* load. What is the average current in each diode?
25. Design a voltage-doubler that produces an average voltage of 250 v across a 10-ohm resistor. The doubler operates from a 60-Hz, 120-v line.
26. What is the ripple factor for the voltage-doubler in Problem 25?
27. Design a voltage-tripler that produces an average voltage of 370 v across a 20-ohm resistor. The tripler operates from a 60-Hz, 120-v line.
28. Calculate the regulation for the voltage-doubler of Problem 25 if the no-load voltage is considered to be $2E_m$. Repeat for the tripler of Problem 27 using $3E_m$ as E_{NL}.
29. For the rectifier circuit shown, calculate E_{rms}, E_{av}, $E_{a\text{-}c}$, ripple factor, power in the load, and the PIV on the diodes.

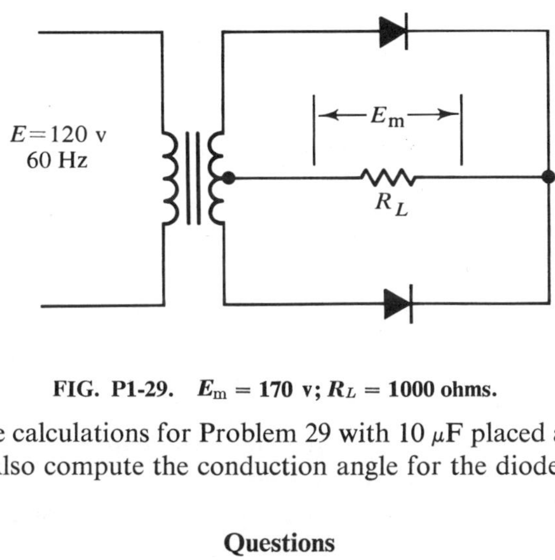

FIG. P1-29. $E_m = 170$ **v;** $R_L = 1000$ **ohms.**

30. Repeat the calculations for Problem 29 with 10 μF placed across the load resistor. Also compute the conduction angle for the diodes.

Questions

1. Define the voltage components that make up the output voltage waveform of a half-wave rectifier.
2. How is the form factor of the output wave of a rectifier circuit affected by load changes?
3. In a half-wave rectifier is a-c power taken from the source during the nonconducting portion of the cycle? Why?

4. Define peak-inverse voltage.
5. Why is the average d-c voltage increased by adding capacitance across the resistive load in a rectifier circuit?
6. Why does the effective value of the a-c component in the output voltage of a rectifier decrease as the capacitance across the load increases?
7. What determines the peak diode current in a half-wave rectifier with an *RC* load?
8. Explain the operation of a half-wave rectifier. Use circuit diagrams and waveforms in your explanation.
9. Why is a silicon diode a more efficient rectifying element than a selenium diode?
10. How are silicon diodes produced?
11. What factors are of primary importance in a rectifying circuit?
12. Define ripple and ripple factor.
13. Define average voltage and rms voltage.
14. Derive the expressions for average and rms voltages of a half-wave rectifier.
15. Derive the expressions for average and rms voltages from a full-wave rectifier.
16. Derive the expressions for average power in half- and full-wave rectifiers.
17. If a diode is placed in series with the load in a 60-Hz circuit, the power loss in the load is reduced. (Assume a perfect diode.) Why is the power reduced? Why does the efficiency of the circuit remain the same. Explain your answers.
18. Define form factor.
19. Draw the circuit for a half-wave rectifier with an *RC* load. Show the direction of current in the load resistor and in the capacitor from 0 to $\pi/2$, $\pi/2$ to π, π to 2π.
20. Repeat question 19 for a full-wave rectifier.
21. What is the PIV across the diode in a half-wave rectifier with an *RC* load? Draw the circuit and indicate the point where the PIV occurs.
22. Repeat Question 21 for a full-wave rectifier.
23. What factors determine the average value of current in a rectifier circuit with an *RC* load? Why?
24. How will the ripple factor of a half-wave rectifier circuit vary as the *RC* time constant of the load is increased?
25. A half-wave rectifier is supplying current to an *RC* load. At what point in the cycle does the diode conduct? Why? At what point does the diode stop conducting? Why?
26. At what point in the cycle does the peak current through a diode occur when a rectifier is supplying an *RC* load?

27. How will a variation in input frequency affect the peak current in a diode that is supplying current to an *RC* load?
28. An *RC* load is connected to the output terminals of a half-wave rectifier, and the angles at which the diode starts and ceases to conduct are measured. How will these angles change if the same load is connected to a full-wave rectifier?
29. Why does the ripple factor decrease if an *RC* load is removed from a half-wave rectifier and connected to a full-wave rectifier?
30. After drawing a vector diagram, show by use of vectors the quadrants where conduction starts and ends when a rectifier is supplying an *RC* load.
31. Using vectors, explain the following equation for a rectifier with an *RC* load.

$$\sin \omega t_1 = \frac{E_m - \Delta E}{E_m}$$

32. If the value of resistance in an *RC* load connected to a rectifier circuit is reduced, how are the following factors affected: ripple factor, average current, and maximum diode current?
33. How are the average voltage, PIV, and ripple factor affected when inductance is added in series with the resistive load of a rectifier?
34. Why does the conduction angle of the diode in a half-wave rectifier increase as inductance is added in series with a resistive load?
35. What is the PIV in terms of E_m for the following rectifier circuits: half-wave; full-wave bridge; full-wave, center-tapped, transformer circuit; half-wave with *RC* load; half-wave with *RL* load; full-wave bridge with *RC* load; and full-wave, center-tapped transformer circuit with *RC* load?
36. A full-wave bridge with a maximum PIV rating of 50 v is used to supply d-c power from a 120-v line. A series resistor is used to reduce the voltage across the diodes. Explain what would happen if the load were to be removed from the output terminals of the bridge.
37. The diode in a half-wave rectifier circuit is supplying current to an *RC* load, and the maximum peak current is being passed by the diode. Should the value of the load resistor be increased or decreased in order to reduce the peak current? Why?
38. What is the peak-inverse voltage across diode 1 and diode 2 in the doubler circuit shown in Fig. 1-16?
39. What limits the amount of power that may be taken from a doubler circuit?
40. Draw the circuit for a voltage-quadrupler.
41. What is the PIV on a half-wave rectifier supplying power through a π filter with capacitance input?

Polyphase Rectifiers

2

The greatest percentage of rectifier circuits is of the single-phase variety. However, in applications requiring large amounts of power and also low-ripple factors, large-filter networks are necessary. In such cases the polyphase rectifier is generally used.

2-1 THREE-PHASE, STAR-RECTIFIER WITH A RESISTIVE LOAD

The simplest polyphase rectifier is the three-phase, half-wave rectifier with star-connected transformer secondary, shown in Fig. 2-1. In the discussion that follows, the effects of transformer resistance, leakage, and current overlap are neglected. Furthermore, it is assumed that the internal diode drops are small.

Owing to the unidirectional characteristics of the diodes, only the transformer secondary winding providing the highest instantaneous voltage can remain connected to the load circuit. From an inspection of the output-waveform shown in Fig. 2-1(b), it is noted that phase A has the highest instantaneous positive voltage from zero to $2\pi/3$ radians. During this time diode 1 is conducting and point D is, for practical purposes, connected directly to point A because of the very low forward resistance of diode 1. Diodes 2 and 3 are reverse-biased during this period, and diode 2 cannot conduct until the voltage of phase A is less than that of phase B. This event occurs just beyond $2\pi/3$ radians, at which time diode 2 conducts and diodes 1 and 3 are reverse-biased. At $4\pi/3$ diode 3 conducts. Diodes 1 and 2 are cut off from $4\pi/3$ to 2π radians. At 2π radians the cycle repeats. The rectified load voltage follows along the tops of the individual voltage waves, as shown in Fig. 2-1(b).

The three-phase half-wave rectifier is superior to the half- and full-wave single-phase rectifiers because of the higher-average d-c voltage produced. The

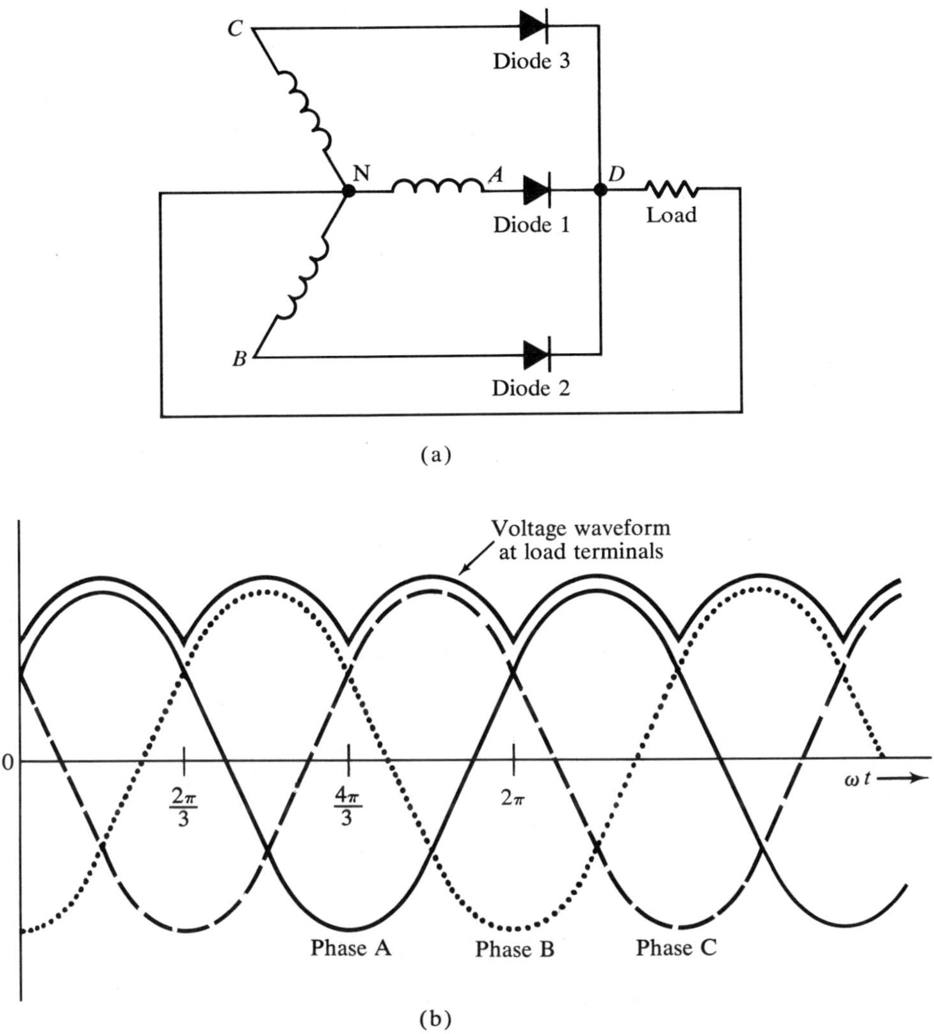

FIG. 2-1. (a) Three-phase half-wave rectifier; (b) Phase- and load-voltages.

three-phase rectifier also has a lower-ripple voltage and a higher-ripple frequency than occur with the single-phase rectifier. The diodes are subjected to the lower peak-inverse voltage of $E_m\sqrt{3}$ rather than of $2E_m$ as in the full-wave, single-phase, center-tapped rectifier (where E_m is the line-to-neutral voltage for a three-phase, star-connected rectifier).

In calculating the voltage and current relationships in a polyphase circuit, it is customary to make certain assumptions; as the number of phases and of diodes increase, the error resulting from these assumptions decreases. We shall assume that the diode voltage drops are constant, the supply transformers have zero reactance, and no overlapping of conduction angles for the diodes occurs.

Each diode conducts for a period equal to $2\pi/N$ where N is the number of conducting diodes. (This period does not apply to the single-phase, half-wave rectifier.) The expression for the conduction angle can be derived by use of Fig. 2-2.

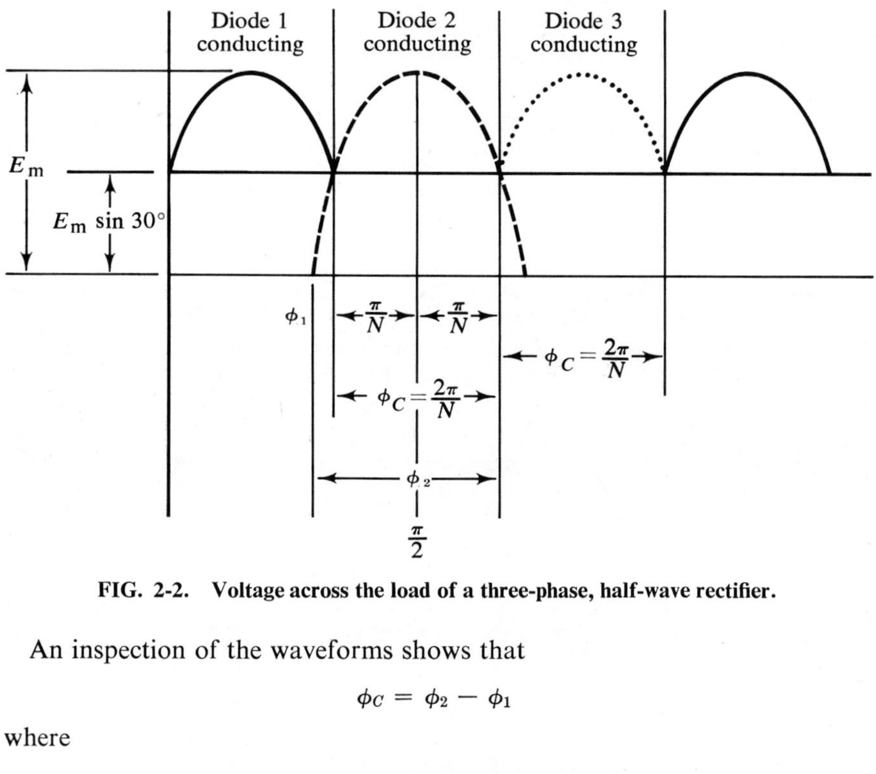

FIG. 2-2. Voltage across the load of a three-phase, half-wave rectifier.

An inspection of the waveforms shows that

$$\phi_C = \phi_2 - \phi_1$$

where

$$\phi_1 = \frac{\pi}{2} - \frac{\pi}{N} \quad \textbf{(2-1)}$$

and

$$\phi_2 = \phi_1 + \frac{2\pi}{N}$$

Therefore, the conduction angle ϕ_C for each diode becomes

$$\phi_C = \phi_1 + \frac{2\pi}{N} - \phi_1$$

or

$$\phi_C = \frac{2\pi}{N} \tag{2-2}$$

The average voltage can now be found by direct integration.

$$E_{\text{d-c}} = \frac{1}{\phi_2 - \phi_1} \int_{\phi_1}^{\phi_2} E_{\text{m}} \sin \omega t (d\omega t)$$

or

$$E_{\text{d-c}} = \frac{N}{2\pi} \int_{\frac{\pi}{2}-\frac{\pi}{N}}^{\frac{\pi}{2}+\frac{\pi}{N}} E_{\text{m}} \sin \omega t (d\omega t) \tag{2-3}$$

Equation 2-3 reduces to $E_{\text{d-c}} = \frac{N}{\pi}(E_{\text{m}})\left(\sin \frac{\pi}{N}\right)$, where E_{m} is equal to the line-to-neutral voltage in all half-wave polyphase bridge circuits.

Example 2-1 As an exercise in the use of Eqs. 2-2 and 2-3, the conduction angle for each diode and the average d-c voltage in a single-phase, full-wave, center-tapped bridge will be calculated.

From Eq. 2-1

$$\phi_1 = \frac{\pi}{2} - \frac{\pi}{2}$$

or

$$\phi_1 = 0$$

$$\phi_2 = \phi_1 + \frac{2\pi}{n}$$

and, if $n = 2$,

$$\phi_2 = \pi$$

The expression for the conduction angle is

$$\phi_C = \phi_2 - \phi_1$$

If $\phi_1 = 0$ and $\phi_2 = \pi$, then

$$\phi_C = \pi$$

From Eq. 2-3

$$E_{\text{d-c}} = \frac{N}{2\pi} \int_{\frac{\pi}{2}-\frac{\pi}{N}}^{\frac{\pi}{2}+\frac{\pi}{N}} E_{\text{m}} \sin \omega t (d\omega t)$$

or

$$E_{\text{d-c}} = \frac{2}{2\pi}\int_{\frac{\pi}{2}-\frac{\pi}{2}}^{\frac{\pi}{2}+\frac{\pi}{2}} E_{\text{m}} \sin \omega t(d\omega t)$$

This expression reduces to

$$E_{\text{d-c}} = \frac{1}{\pi}\int_{0}^{\pi} E_{\text{m}} \sin \omega t(d\omega t)$$

or

$$E_{\text{d-c}} = \frac{2E_{\text{m}}}{\pi}$$

This equation is the same as that for the average voltage from a full-wave bridge, as derived in Chapter 1.

The three-phase, half-wave rectifier will next be analyzed. In referring to the circuit of Fig. 2-1, it can be seen that the average voltage can be determined by the use of Eqs. 2-2 and 2-3.

$$E_{\text{d-c}} = \frac{3}{2\pi}\int_{\frac{\pi}{6}}^{\frac{5\pi}{6}} E_{\text{m}} \sin \omega t(d\omega t)$$

or

$$E_{\text{d-c}} = \frac{3E_{\text{m}}}{2\pi}\Big[-\cos \omega t\Big]_{30°}^{150°} \tag{2-4}$$

which reduces to

$$E_{\text{d-c}} = 0.827\ E_{\text{m}} \tag{2-5}$$

The rms (effective) value of the load voltage can be found by integration as follows:

$$E_{\text{rms}} = \sqrt{\frac{3}{2\pi}\int_{\frac{\pi}{6}}^{\frac{5\pi}{6}} (E_{\text{m}})^2 \sin^2 \omega t(d\omega t)} \tag{2-6}$$

$$E_{\text{rms}} = \sqrt{\frac{3E_{\text{m}}{}^2}{2\pi}\left(\frac{\omega t}{2} - \frac{\sin \omega t \cos \omega t}{2}\right)_{\frac{\pi}{6}}^{\frac{5\pi}{6}}} \tag{2-7}$$

$$E_{\text{rms}} = 0.838\ E_{\text{m}} \tag{2-8}$$

The ripple factor is then computed from

$$\gamma = \frac{E_{\text{a-c}}}{E_{\text{d-c}}}$$

To obtain $E_{\text{a-c}}$,

$$E_{\text{a-c}} = \sqrt{(E_{\text{rms}})^2 - (E_{\text{d-c}})^2}$$

From Eqs. 2-5 and 2-8

$$E_{\text{a-c}} = \sqrt{(0.838\ E_{\text{m}})^2 - (0.827\ E_{\text{m}})^2}$$

which reduces to

$$E_{\text{a-c}} = 0.140\ E_{\text{m}} \tag{2-9}$$

Then,

$$\gamma = \frac{0.140\ E_{\text{m}}}{0.827\ E_{\text{m}}} \times 100 = 17 \text{ percent} \tag{2-10}$$

The peak-inverse voltage on each diode is the line-to-neutral voltage multiplied by $\sqrt{3}$, and it occurs 120° after the phase voltage passes through zero.

Before specifying the size of transformer needed, the rms value of the phase currents in each coil must be found, as well as the volt-ampere rating of each secondary coil.

$$I_{\text{rms}}/\text{phase} = \sqrt{\frac{1}{2\pi}\int_{\frac{\pi}{6}}^{\frac{5\pi}{6}} (I_{\text{m}})^2 \sin^2 \omega t(d\omega t)} \tag{2-11}$$

$$I_{\text{rms}}/\text{phase} = 0.486\ I_{\text{m}}$$

Note that this rms value is averaged over $2\ \pi$ radians or one full cycle. The volt-ampere rating of each secondary coil is next found.

$$\text{vA}_{\text{s}} = \frac{E_{\text{m}}}{\sqrt{2}}(I_{\text{rms}}/\text{phase}) \tag{2-12}$$

The total vA rating of the secondaries is three times the value determined by Eq. 2-12. For a three-phase, half-wave circuit, therefore,

$$\text{Total vA}_{\text{s}} = \frac{3}{\sqrt{2}} E_{\text{m}} \left(\frac{I_{\text{rms}}}{\text{phase}}\right) \tag{2-13}$$

In order to relate the d-c power output to the required transformer rating, the d-c output must be expressed in terms of the a-c input. From Eq. 2-5

$$E_{\text{d-c}} = 0.827\ E_{\text{m}}$$

and

$$I_{\text{d-c}} = 0.827\ I_{\text{m}}$$

From Eq. 2-11 $I_{rms}/\text{phase} = 0.486\, I_m$

Computing from Eqs. 2-5 and 2-11

$$I_{rms}/\text{phase} = \frac{0.486}{0.827} I_{d\text{-}c}$$

Substituting these expressions for current and voltage into Eq. 2-13, we have

$$vA_s = \frac{3}{\sqrt{2}}\left(\frac{E_{d\text{-}c}}{0.827}\right)\left(\frac{0.486}{0.827}\right) I_{d\text{-}c}$$

which reduces to $vA_s = 1.5\, I_{d\text{-}c} E_{d\text{-}c}$

The volt-ampere rating of a transformer delivering power to a three-phase, half-wave rectifier must, therefore, be 1.5 times the volt-ampere rating of the load.

The rms current in the three-phase, Y-connected load is $\sqrt{3}\,(I_{rms})$ per phase, or, from Eq. 2-11

$$I_{rms} = \sqrt{3}\,(0.486) I_m$$

$$I_{rms} = 0.84\, I_m \qquad \textbf{(2-14)}$$

The average power in the load may now be computed as follows:

$$P_{av} = (0.84\, I_m)^2 R$$

$$P_{av} = 0.709\, (I_m)^2 R \qquad \textbf{(2-15)}$$

or

$$P_{av} = \frac{0.709\, (E_m)^2}{R} \qquad \textbf{(2-16)}$$

Example 2-2 A three-phase, half-wave rectifier is to deliver 60 A to a load at 250 v (average values). Calculate the:

1. average current per diode
2. PIV per diode
3. volt-ampere rating of the transformer primary
4. voltage rating of the secondary winding
5. rms current in each secondary winding and
6. magnitude of the ripple voltage

1. The average current per diode is $60/\sqrt{3}$ or 34.7 A.
2. The peak-inverse voltage is $E_m\sqrt{3}$. To find E_m use Eq. 2-5.

$$E_{d\text{-}c} = 0.827\, E_m$$

or

$$E_m = 250/0.827 = 300 \text{ v}$$

and

$$\text{PIV} = 520 \text{ v}$$

3. The volt-ampere rating for the transformer winding is found from the average values of E and I. From Eq. 2-13 the volt-ampere rating for the transformer becomes

$$\text{vA}_s = 1.5\, E_{\text{d-c}} I_{\text{d-c}}$$

$$\text{vA}_s = 1.5(250)(60) = 22.5 \text{ kvA}$$

4. The rms voltage rating of the secondary winding is found from

$$E_{\text{rms sec}} = E_m/\sqrt{2}$$

$$E_{\text{rms sec}} = 212 \text{ v}$$

5. The rms current per transformer coil is found from Eqs. 2-11 and 2-5.

$$I_{\text{rms per coil}} = 0.486\, I_m$$

From Eq. 2-5

$$I_{\text{d-c}} = 0.827\, I_m$$

$$I_{\text{rms per coil}} = I_{\text{d-c}} \frac{0.486}{0.827} = 60(0.59) = 35 \text{ A}$$

6. The magnitude of the ripple voltage may be found from the definition of ripple voltage and the knowledge that the three phase, half-wave rectifier produces a ripple of 17 percent (Eq. 2-10). Therefore,

$$0.17 = \frac{E_{\text{a-c}}}{E_{\text{d-c}}}$$

$$0.17(250) = E_{\text{a-c}}$$

$$E_{\text{a-c}} = 42.5 \text{ v}$$

2-2 THREE-PHASE BRIDGE RECTIFIER

One of the most effective types of rectifying circuits from the standpoint of ripple factor, peak-inverse voltage, and *transformer utilization* is the three-phase bridge rectifier shown in Fig. 2-3. This bridge circuit can be considered as two single, half-wave circuits connected *back-to-back*. Like the full-wave, single-phase bridge, this bridge circuit conducts current to the load during both halves of each cycle. The ripple voltage is very low and is of the same magnitude as that from a six-phase, half-wave circuit. Filtering is generally not required for most applications. The d-c output voltage of the three-phase, full-wave rectifier is greater than that of the three-phase half-wave circuit, and the volt-ampere rating of the transformer secondary is reduced to 1.05 $I_{\text{d-c}}\, E_{\text{d-c}}$.

To analyze the operation of this circuit, refer to Fig. 2-3. The diodes always

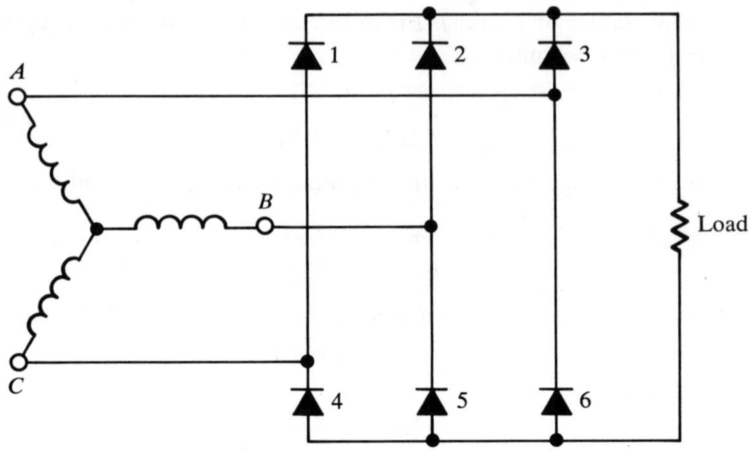

(a)

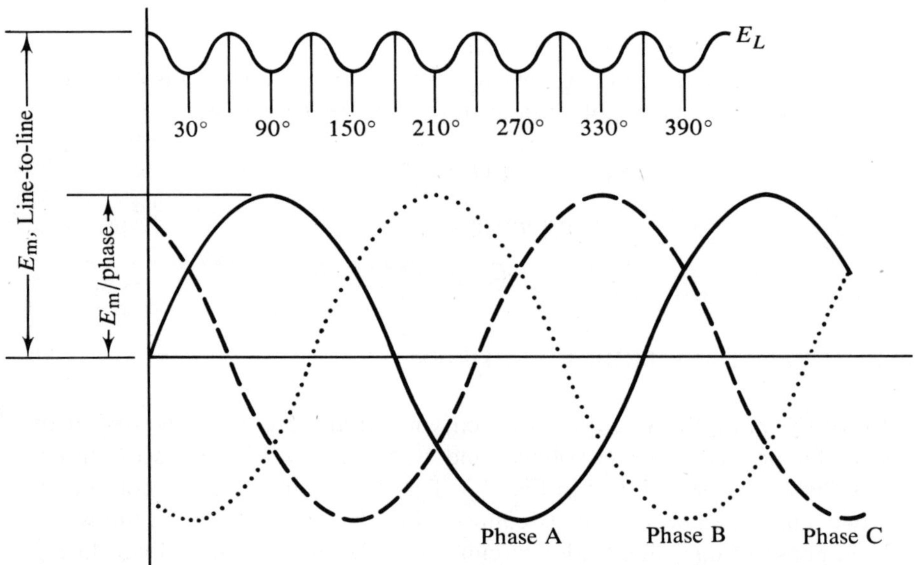

(b)

FIG. 2-3. (a) Three-phase, full-wave rectifier; (b) Phase voltages and load voltage.

conduct between the line-to-line voltages between which the greatest potential difference occurs.

At $30° < \omega t < 90°$ the greatest potential exists between phases A and B, with A positive. Diodes 3 and 5 conduct.

At $90° < \omega t < 150°$ the greatest potential exists between phases A and C, with phase A positive. Diodes 3 and 4 conduct.

At $150° < \omega t < 210°$ the greatest potential exists between phases B and C, with B positive. Diodes 2 and 4 conduct.

At $210° < \omega t < 270°$ the greatest potential difference exists between phases B and A, with phase B positive. Diodes 2 and 6 conduct.

At $270° < \omega t < 330°$ the greatest potential exists between phases C and A, with C positive, and diodes 1 and 6 conduct.

At $330° < \omega t < 390°$ the greatest potential exists between phases C and B, with C positive, and diodes 1 and 5 conduct.

Each *combination* of two diodes conducts for 60°, and the cycle then repeats.

A graphic approach to the analysis of the three-phase, full-wave bridge will be used to acquaint the reader with this method of analysis. Although graphic analysis introduces some error into the calculations of $E_{\text{d-c}}$, E_{rms}, and similar values, such an approach leads to a clear understanding of the waveforms involved. The waveform of a three-phase, full-wave rectifier is shown in Fig. 2-4. Assume that this waveform consists of a series of pulses having triangular peaks and a constant base voltage E_c. The average value of the wave is equal to E_c plus the average value of the triangular waveform E_T. The angle ϕ_1 may be calculated from Eq. 2-1.

$$\phi_1 = \frac{\pi}{3}$$

For all polyphase bridge calculations, E_{m} is the line to line voltage. The d-c voltage is

$$E_{\text{d-c}} = E_c + E_{T_{\text{av}}} \tag{2-17}$$

where

$$E_c = E_{\text{m}} \sin \frac{\pi}{3}$$

From which

$$E_c = 0.866\ E_{\text{m}} \tag{2-18}$$

The peak-to-peak value of the a-c component is $E_T = E_{\text{m}} - E_c$. From Eq. 2-18,

$$E_c = 0.866\ E_{\text{m}}$$

Therefore,

$$E_T = E_{\text{m}}[1 - 0.866]$$

$$E_T = 0.134\ E_{\text{m}}$$

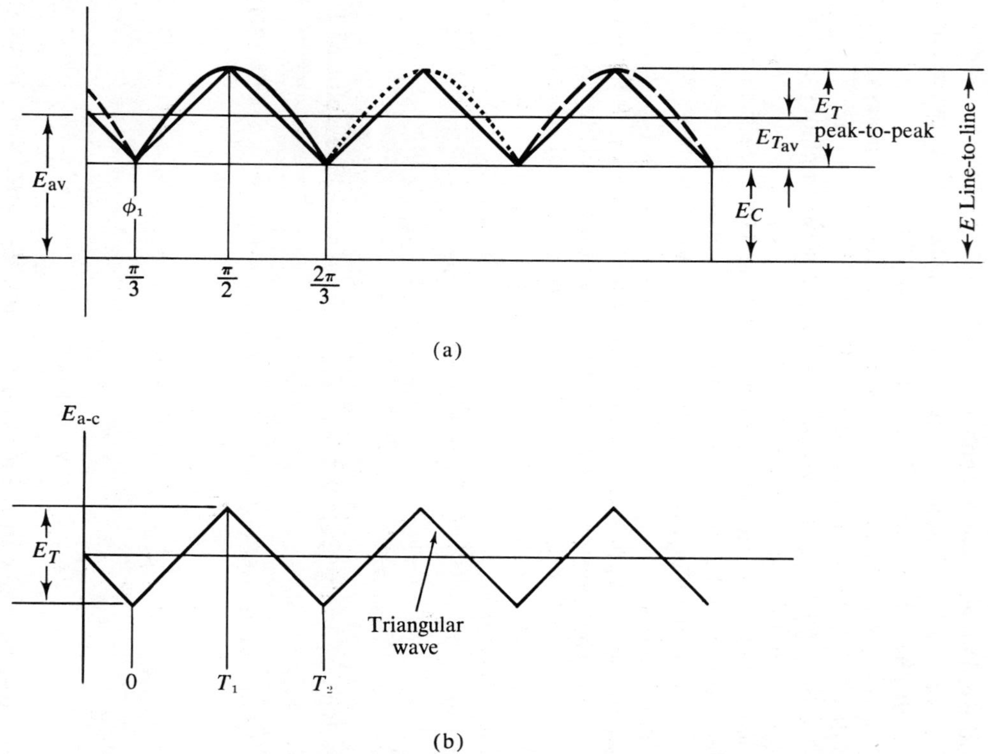

FIG. 2-4. (a) Triangular representation of three-phase, full-wave rectifier voltage waveforms; (b) Waveform of assumed triangular-ripple voltage.

The waveform of the a-c component is assumed to be triangular, and its average value can be expressed as

$$E_{T_{av}} = E_T/2$$

or

$$E_{T_{av}} = \frac{0.134\ E_{\mathrm{m}}}{2}$$

from which

$$E_{T_{av}} = 0.067\ E_{\mathrm{m}} \tag{2-19}$$

From Eq. 2-17,

$$E_{\mathrm{d\text{-}c}} = E_c + E_{T_{av}}$$

Substituting into this equation the expressions for E_c and $E_{T_{av}}$ from Eqs. 2-18 and 2-19, we have

$$E_{\mathrm{d\text{-}c}} = 0.866\ E_{\mathrm{m}} + 0.067\ E_{\mathrm{m}}$$

which reduces to

$$E_{\mathrm{d\text{-}c}} = 0.933\ E_{\mathrm{m}} \tag{2-20}$$

If the technique of integration were used to find $E_{\mathrm{d\text{-}c}}$, then

$$E_{\mathrm{d\text{-}c}} = 0.953\ E_{\mathrm{m}}$$

Thus, the assumption of a triangular waveshape introduces an error of less than 3 percent.

The value of the ripple voltage $E_{\mathrm{a\text{-}c}}$ can be determined by finding the effective value for the triangular waveform shown in Fig. 2-4(b). By integration, the rms (effective) value of the triangular waveform is

$$E_{\mathrm{a\text{-}c}} = \frac{E_T}{2\sqrt{3}}$$

Substituting

$$E_T = 0.134\ E_{\mathrm{m}}$$

$$E_{\mathrm{a\text{-}c}} = \frac{0.134\ E_{\mathrm{m}}}{2\sqrt{3}}$$

or

$$E_{\mathrm{a\text{-}c}} = 0.039\ E_{\mathrm{m}} \tag{2-21}$$

To find the rms value of the voltage at the load, Eq. 1-17 is used.

$$E_{\mathrm{rms}} = \sqrt{(E_{\mathrm{d\text{-}c}})^2 + (E_{\mathrm{a\text{-}c}})^2}$$

or

$$E_{rms} = \sqrt{(0.933\ E_m)^2 + (0.039\ E_m)^2}$$

which reduces to

$$E_{rms} = 0.935\ E_m \tag{2-22}$$

The technique of integration would give a value of $E_{rms} = 0.955\ E_m$; this approximate solution has an error of less than 3 percent. The ripple factor can be calculated as follows:

$$\gamma = E_{a\text{-}c}/E_{d\text{-}c}$$

$$\gamma = \frac{0.039\ E_m}{0.933\ E_m} \times 100 = 4.2 \text{ percent} \tag{2-23}$$

Using the values obtained through calculus, the ripple factor would be approximately 4 percent.

The frequency of the ripple is six times the frequency of the input voltage and is filtered easily. The peak-inverse voltage is equal to $\sqrt{3}E_m$.

The average power is expressed as

$$P_{av} = (I_{rms})^2 R$$

Then, from Eq. 2-20

$$P_{av} = \frac{(0.933\ E_m)^2}{R} \tag{2-24}$$

or

$$P_{av} \cong E_{d\text{-}c} I_{d\text{-}c} \tag{2-25}$$

The average current carried by each diode is one-third of the average d-c current.

$$I_{d\text{-}c} \text{ per diode} = \frac{0.933\ E_m}{3\ R_L} = \frac{0.311\ E_m}{R_L} \tag{2-26}$$

Example 2-3 A three-phase, full-wave rectifier is to deliver 60 A to a load at 250 v (average values). Calculate the:

1. average current per diode
2. PIV per diode
3. vA rating of the transformer primary
4. voltage rating of the secondary winding
5. rms current in each winding
6. magnitude of the ripple voltage

1. The average current per diode is

$$I_{av}/3$$

Therefore, $$I_{av} = \frac{60}{3} = 20 \text{ A}$$

2. The peak-inverse voltage per diode is $E_m\sqrt{3}$, and E_m may be found from Eq 2-20.

$$E_{d\text{-}c} = 0.933\, E_m$$

or

$$E_m = 250 \text{ v}/0.933$$

$$E_m = 268 \text{ v (line-to-line voltage)}$$

from which

$$\text{PIV} = 268\sqrt{3}$$

or

$$\text{PIV} = 465 \text{ v}$$

3. The volt-ampere rating for the transformer is found from the rms values of E and I.

$$\text{vA}_s = \frac{(E_{rms})^2}{R_L}$$

$$R_L = 250 \text{ v}/60 \text{ A} = 4.15 \text{ ohms}$$

From Eq. 2-22

$$E_{rms} = 0.935\, E_m$$

Therefore,

$$\text{vA}_s = \frac{(0.935\, E_m)^2}{4.15} = \frac{0.87\, E_m{}^2}{4.15}$$

Then, if

$$E_m = 268 \text{ v}$$

$$\text{vA}_s = \frac{0.87(268)^2}{4.15} = 14.5 \text{ kvA}$$

4. The voltage rating of each secondary winding is found by

$$E_m = E_{rms}\sqrt{2}$$

then

$$E_{rms} = \frac{268}{\sqrt{2}}$$

or

$$E_{rms} = 190 \text{ v}$$

which is the rms value of line-to-line voltage. The rms voltage from each winding is $E_{rms}/\text{winding} = 190 \text{ v}/\sqrt{3} = 110$ v.

5. The rms current in each winding is $1/\sqrt{3}$ of the rms current in the load.

$$\underset{\text{(load)}}{I_{rms}} = \frac{0.935\, E_m}{4.15\, \Omega}$$

$$\underset{\text{(load)}}{I_{rms}} = 60.6 \text{ A}$$

$$I_{rms}/\text{secondary winding} = \frac{60.6}{\sqrt{3}} = 34.8 \text{ A}$$

6. The ripple voltage from Eq. 2-22 is

$$E_{ripple} = 4.2 \text{ percent } E_{d\text{-}c}$$
$$= 0.042(250) \text{ v}$$
$$E_{ripple} = 10.5 \text{ v}$$

2-3 TRANSFORMER UTILIZATION FACTOR

Transformers used in rectifier circuits do not have sinusoidal currents through both the primary and secondary windings. The reason is easy to understand if the half-wave, single-phase rectifier is used as an example. In this circuit the secondary current continues for only one-half cycle and is not sinusoidal, but current in the primary flows for the full cycle. The I^2R heating effects, therefore, are not equal for the primary and secondary windings because of the difference in the average values of current. Since the primary and secondary currents have different waveforms, a different factor must be used in determining the rating of each winding. Transformers are generally rated in volt-amperes rather than in power delivered to the load, because the heat rise in the windings is determined by the volt-amperes delivered. This volt-ampere rating is based on a sinusoidal current through the windings.

The *utility factor*, also called the *utilization factor* of a transformer, is the ratio of the rectified power output to the volt-ampere rating of the winding. The ratio may be between the d-c power output and the volt-amperes of the secondary windings. This ratio is otherwise referred to as the *secondary utilization factor*. A ratio may also express the relationship between the d-c power output and the volt-amperes of the primary windings. The second ratio is more commonly referred to as the *primary utilization factor*. It will be shown that the primary and secondary utilization factors are not necessarily equal.

The utilization factor, abbreviated UF, can be expressed as

$$\mathrm{UF} = \frac{\text{total d-c power}}{\text{total volt-ampere rating of the winding}} \tag{2-27}$$

From Eq. 2-27 it is apparent that the maximum utilization factor is unity.

For the case of a single-phase, half-wave rectifier the primary and secondary current waveforms are shown in Fig. 2-5. In Fig. 2-5(b) it can be noted that current in the secondary winding lasts for only one-half cycle. The expressions for the load current and load voltage are

$$I_{\text{d-c}} = \frac{I_m}{\pi}$$

and

$$E_{\text{d-c}} = \frac{E_m}{\pi}$$

The d-c power delivered to the load is

$$P_{\text{d-c}} = E_{\text{d-c}} I_{\text{d-c}}$$

or

$$P_{\text{d-c}} = \frac{1}{\pi^2} E_m I_m$$

This expression reduces to

$$P_{\text{d-c}} = 0.101\ E_m I_m \tag{2-28}$$

Before determining the utility factor of this transformer, it is necessary to find the value of a sinusoidal current that would have the same heating effect on the secondary winding as a pulsating direct current. Assume that the voltage rating of the secondary is unchanged. The rms value of load current and of current in the secondary winding is

$$I_{\text{rms}} = \frac{I_m}{2}$$

This is also the rms value of an equivalent sinusoidal current which will produce heat in this winding. The volt-ampere rating of a transformer delivering an equal sinusoidal current is

$$\text{vA} = \frac{E_m}{\sqrt{2}} \times \frac{I_m}{2} = 0.354\ E_m I_m$$

By definition, the utility factor of a transformer is

$$\mathrm{UF} = \frac{P_{\text{d-c}}}{\text{vA}}$$

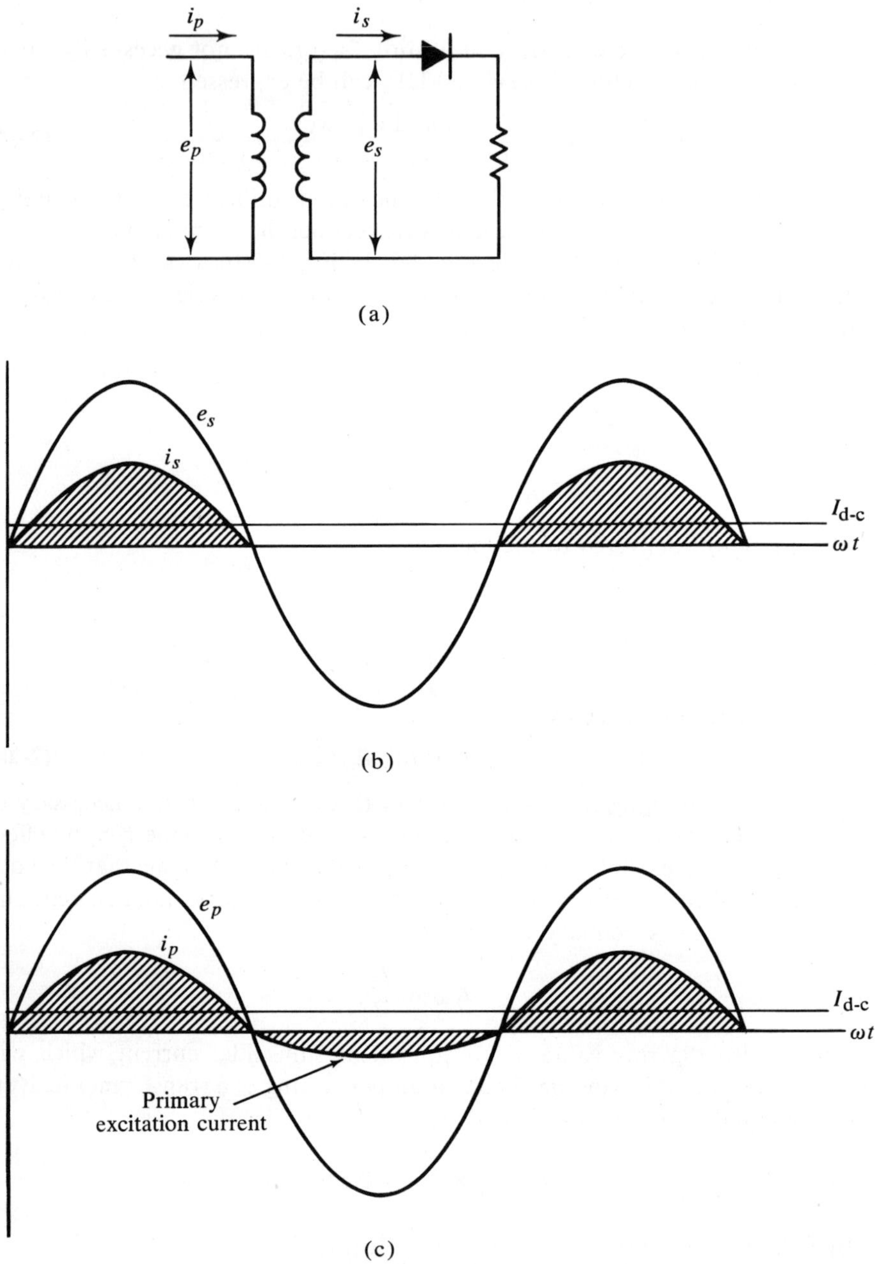

FIG. 2-5. (a) Half-wave rectifier circuit; (b) Secondary voltage and current waveforms; (c) Primary voltage and current waveforms.

For the half-wave rectifier,

$$\mathrm{UF_s} = \frac{0.101\, E_m I_m}{0.354\, E_m I_m}$$

$$\mathrm{UF_s} = 0.285$$

Thus the secondary winding of a transformer used in a half-wave rectifier circuit must be 1/UF (3.49) times larger than the secondary of a transformer that is required to deliver the same amount of a-c power.

Direct current does not pass through the primary windings, and the a-c current is nearly sinusoidal. However, the unidirectional current in the secondary tends to saturate the iron core of the transformer, and the primary excitation current is greater than it would be if an a-c load were connected to the secondary. The utilization factor of the primary is, therefore, less than 1, but it is higher than that of the secondary UF. A typical primary utilization factor is 0.38 for a half-wave rectifier.

Transformers designed for use in rectifier circuits generally are wound with larger sizes of wire than the sizes used with equivalent a-c power transformers; compensation for the higher losses is made by this means. Choke-input filters, which sustain current over more than one-half cycle, raise the utility factor above that of a pure resistive load. For a full-wave rectifier with choke-input filter utility factors of 0.637 for the secondary and 0.9 for the primary are possible.

For a half-wave rectifier of N phases using N diodes the secondary utilization factor may be derived as follows:

$$\mathrm{UF_s} = \frac{P_{\text{d-c}}}{\mathrm{vA}}$$

and

$$P_{\text{d-c}} = E_{\text{d-c}} I_{\text{d-c}}$$

Using Eq. 2-3

$$E_{\text{d-c}} = \frac{NE_m}{2\pi} \int_{\frac{\pi}{2}-\frac{\pi}{N}}^{\frac{\pi}{2}+\frac{\pi}{N}} \sin \omega t (d\omega t) \qquad \text{(Eq. 2-3)}$$

This reduces to

$$E_{\text{d-c}} = E_m \frac{N}{\pi} \sin \frac{\pi}{N}$$

$$I_{\text{d-c}} = \frac{E_{\text{d-c}}}{R_L}$$

So that

$$P_{\text{d-c}} = \frac{(E_m)^2}{R_L}\left(\frac{N}{\pi}\right)^2 \sin^2\frac{\pi}{N}$$

Before the volt-ampere rating of the secondary windings can be found, the rms value of the secondary current in each winding must first be calculated. From Eq. 2-11

$$I_{\text{rms}} = \sqrt{\frac{1}{2\pi}\int_{\frac{\pi}{2}-\frac{\pi}{N}}^{\frac{\pi}{2}+\frac{\pi}{N}} \left(\frac{E_m}{R}\right)^2 \sin^2 \omega t(d\omega t)}$$

or

$$I_{\text{rms}} = \frac{E_m}{R}\sqrt{\frac{1}{2\pi}\left(\frac{\pi}{N} + \sin\frac{\pi}{N}\cos\frac{\pi}{N}\right)}$$

The total secondary volt-ampere rating $E_{\text{rms}}I_{\text{rms}}$ may be expressed as

$$\text{vA}_s = \left(\frac{NE_m}{\sqrt{2}}\right)\left(\frac{E_m}{R}\right)\sqrt{\frac{1}{2\pi}\left(\frac{\pi}{N} + \sin\frac{\pi}{N}\cos\frac{\pi}{N}\right)}$$

The UF becomes

$$\text{UF}_s = \frac{\dfrac{(E_m)^2}{R}\left(\dfrac{N}{\pi}\sin^2\dfrac{\pi}{N}\right)}{\left(\dfrac{NE_m}{\sqrt{2}}\right)\left(\dfrac{E_m}{R}\right)\sqrt{\dfrac{1}{2\pi}\left(\dfrac{\pi}{N} + \sin\dfrac{\pi}{N}\cos\dfrac{\pi}{N}\right)}}$$

or

$$\text{UF}_s = \frac{2\dfrac{N}{\pi}\left(\sin^2\dfrac{\pi}{N}\right)}{\sqrt{\pi\left(\dfrac{\pi}{N} + \sin\dfrac{\pi}{N}\cos\dfrac{\pi}{N}\right)}} \tag{2-29}$$

Equation 2-29 is the general expression for the secondary utilization factor of a half-wave rectifier with N diodes supplying a resistive load.

The primary utilization factor can be expressed as

$$\text{UF}_p = \frac{N}{P\sqrt{2}}(\text{UF}_s)$$

where P is the number of primary phases.

If the utilization factor vs the number of phases were plotted in accordance with Eq. 2-29, the curve of Fig. 2-6 would be obtained. It can be easily determined from this curve that maximum utilization occurs when N is approxi-

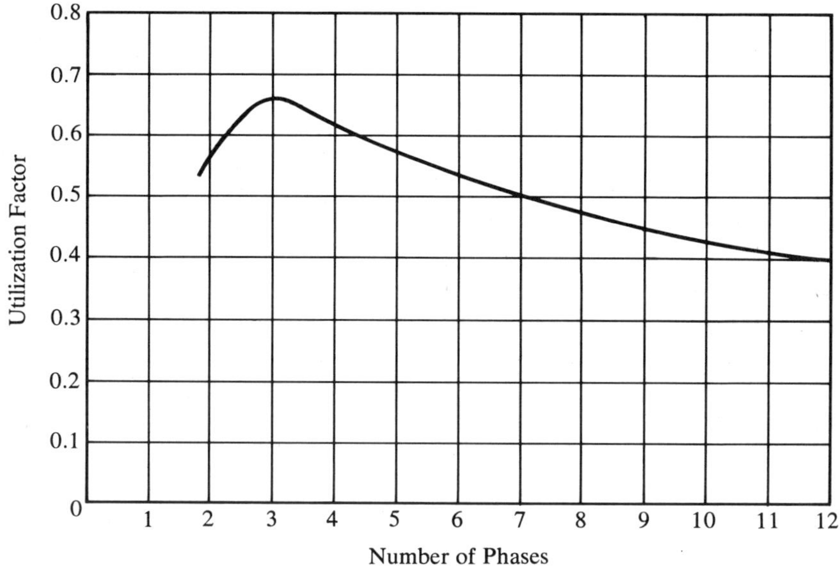

FIG. 2-6. Plot of Eq. 2-29 showing variation in utilization factor as a function of number of phases.

mately 3. Therefore, best transformer utilization is obtained with the three-phase rectifier.

Example 2-4 Using Eq. 2-29, find the secondary utilization factor of the six-phase, half-wave rectifier shown in Fig. 2-8.

$$UF_s = \frac{12(\sin^2 30°)/\pi}{\sqrt{\pi(0.525 + \sin 30° \cos 30°)}}$$

$$UF_s = 0.56$$

2-4 POLYPHASE RECTIFIER WITH AN *RL* LOAD

Table 2-1 indicates that the ripple present in the output of a three-phase rectifier is far below that of a single-phase rectifier, and as the number of phases increases, the ripple is further reduced. In some applications, however, even a small amount of ripple may be objectionable. In such cases an inductor may be connected in series with the load, as shown in Fig. 2-7, to produce the desired amount of filtering.

The inductive loading produces nearly-constant diode-current blocks, as shown in Fig. 2-7(b). This inductive load also produces a constant load cur-

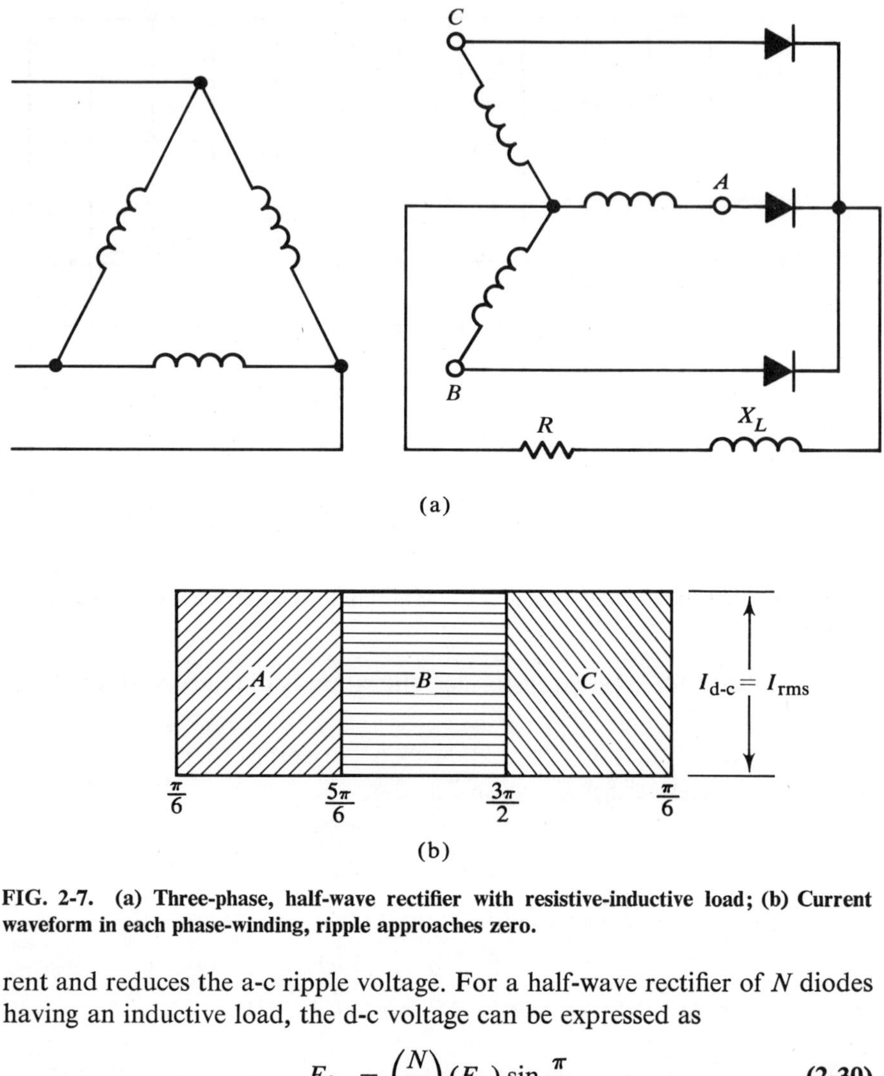

FIG. 2-7. (a) Three-phase, half-wave rectifier with resistive-inductive load; (b) Current waveform in each phase-winding, ripple approaches zero.

rent and reduces the a-c ripple voltage. For a half-wave rectifier of N diodes having an inductive load, the d-c voltage can be expressed as

$$E_{d\text{-}c} = \left(\frac{N}{\pi}\right)(E_m) \sin \frac{\pi}{N} \qquad \textbf{(2-30)}$$

Since the diode current is nearly constant, due to the inductive load, $I_{d\text{-}c} = I_{rms}$, and the power input from the transformer secondaries is

$$P_{a\text{-}c} = I_{rms}\left(\frac{N}{\pi}\right)(E_m) \sin \frac{\pi}{N} \qquad \textbf{(2-31)}$$

When the ripple factor is very low, Eq. 2-31 gives the value of $P_{d\text{-}c}$ (assuming a rectifier efficiency of 100 percent.) The inductor has filtered out all harmonics, and the total output power is usable d-c power.

The rms current per diode is

$$\frac{I_{\text{rms}}}{\text{diode}} = \frac{I_{\text{d-c}}}{\sqrt{N}} \tag{2-32}$$

The a-c volt-ampere output for the transformer secondaries is

$$\text{vA}_s = \frac{E_{\text{m}}}{\sqrt{2}} I_{\text{rms}} = \frac{E_{\text{m}}}{\sqrt{2}} \left(\frac{I_{\text{d-c}}}{\sqrt{N}}\right) N$$

or

$$\text{vA}_s = E_{\text{m}} I_{\text{d-c}} \sqrt{\frac{N}{2}} \tag{2-33}$$

The secondary utilization factor is expressed as

$$\text{UF}_s = \frac{\text{sec power}}{\text{vA}_s} = \frac{I_{\text{d-c}} E_{\text{m}} \left(\frac{N}{\pi} \sin \frac{\pi}{N}\right)}{I_{\text{d-c}} E_{\text{m}} \sqrt{\frac{N}{2}}}$$

or,

$$\text{UF}_s = \frac{\sqrt{2N}}{\pi} \sin \frac{\pi}{N} \tag{2-34}$$

When Eq. 2-34 is maximized, the maximum utility factor is seen to be at $N = 2.7$, or between a two- and a three-phase rectifier.

Example 2-5 A 220-volt, three-phase transformer is supplying power to a three-phase half-wave rectifier, delivering 20 A to a resistive-inductive load. Assuming that the inductance removes the a-c component from the load current, calculate:

1. the d-c power in the load
2. the rms current per winding
3. the vA supplied by the secondary windings
4. the transformer secondary-utilization factor

1. To find the d-c power in the load,

$$P_{\text{d-c}} = E_{\text{d-c}} \times I_{\text{d-c}}$$

$$E_{\text{d-c}} = \frac{N}{\pi} (E_{\text{m}}) \sin \frac{\pi}{N} \qquad \text{(Eq. 2-30)}$$

$$E_{\text{d-c}} = \frac{3}{3.14} (220\sqrt{2}) \sin 60^\circ$$

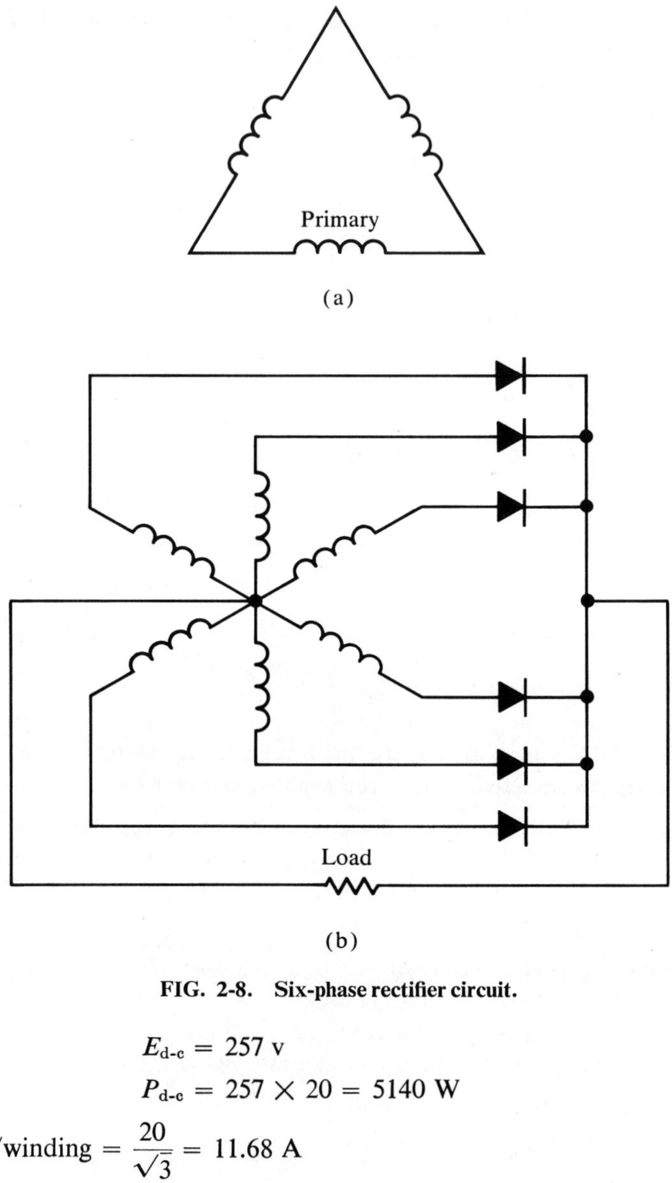

FIG. 2-8. Six-phase rectifier circuit.

$$E_{\text{d-c}} = 257 \text{ v}$$

$$P_{\text{d-c}} = 257 \times 20 = 5140 \text{ W}$$

2. $I_{\text{rms}}/\text{winding} = \dfrac{20}{\sqrt{3}} = 11.68 \text{ A}$

3. The volt-amperes supplied by the secondary windings become

$$\text{vA}_s = E_m I_{\text{d-c}} \sqrt{\frac{N}{2}} \qquad \text{(Eq. 2-33)}$$

$$\text{vA}_s = 220\sqrt{2}(20)\sqrt{\frac{3}{2}}$$

$$\text{vA}_s = 7700$$

4. $$\text{UF}_s = \frac{\text{sec power}}{\text{vA}_s}$$

$$\text{UF}_s = \frac{5140}{7700}$$

$$\text{UF}_s = 0.67$$

The UF_s may also be calculated by using Eq. 2-34.

$$\text{UF}_s = \frac{\sqrt{2}N}{\pi}\left(\sin\frac{\pi}{N}\right)$$

$$\text{UF}_s = \frac{\sqrt{6}}{\pi}(\sin 60°)$$

$$\text{UF}_s = 0.67$$

2-5 SIX-PHASE RECTIFIER

The six-phase rectifier, shown in Fig. 2-8, is especially suitable for use in the mercury-pool type of rectifiers. However, both transformer utilization and output voltage are less than they are for a three-phase bridge rectifier. The values for output voltage and current are given in Table 2-1.

2-6 DOUBLE-WYE RECTIFIER WITH INTERPHASE WINDING

Perhaps the polyphase rectifier most frequently used for supplying large loads is the double-wye rectifier with interphase winding, shown in Fig. 2-9. This circuit, the equivalent of two three-phase half-wave rectifiers connected in parallel, has the advantages of raising the utilization factor of the transformers, providing low-ripple factor at high-frequency low peak-inverse voltage, and high diode-utilization.

The two polyphase transformers are joined together at the neutral point by the interphase winding. Since the two circuits do not produce the same instantaneous output voltages, there is an a-c potential difference at the neutrals, which cannot be tied together directly. The interphase winding reduces the a-c current between the neutrals. However, both secondaries do produce

TABLE 2-1. Comparison of Polyphase Rectifiers

Type of Rectifier	$E_{d\text{-}c}$	PIV	Ripple Factor	vA Secondary	UF Secondary	UF Primary
3ϕ star-connection resistive load	$0.827\ E_m$	$\sqrt{3}\ E_m$	0.187	$1.5\ E_{d\text{-}c}\ I_{d\text{-}c}$	0.66	0.82
3ϕ star-connection resistive-inductive load	$0.827\ E_m$	$\sqrt{3}\ E_m$	less than 0.187*	$1.5\ E_{d\text{-}c}\ I_{d\text{-}c}$	0.66	0.82
3-phase bridge resistive load	$0.953\ E_m$	$\sqrt{3}\ E_m$†	0.06	$1.05\ E_{d\text{-}c}\ I_{d\text{-}c}$	0.95	0.955
3-phase bridge resistive-inductive load	$0.953\ E_m$	$\sqrt{3}\ E_m$†	less than 0.06*	$1.05\ E_{d\text{-}c}\ I_{d\text{-}c}$	0.95	0.955
6-phase star resistive load	$0.96\ E_m$	$2\ E_m$	0.06	$1.81\ E_{d\text{-}c}\ I_{d\text{-}c}$	0.551	0.78
Double-wye with interphase winding; resistive load	$0.827\ E_m$	$\sqrt{3}\ E_m$	less than 0.08*	$1.49\ E_{d\text{-}c}\ I_{d\text{-}c}$	0.67	0.955

* Dependent upon value of L † E_m is the peak value of the line to line voltage

the same *average* output voltage, and there is no d-c difference between the neutrals of the two transformers. At any instant two diodes are conducting, one in each group. Each diode conducts for 120°, which is almost optimum for best transformer utilization. Because of the unequal instantaneous currents in the interphase winding, a voltage of amplitude 2 E_m appears across the winding.

The waveforms for this current are shown in Fig. 2-9(b).

2-7 THREE-PHASE ZIGZAG RECTIFIERS

To avoid the d-c saturation inherent in three-phase half-wave rectifier transformers, the zigzag secondary winding is used. This circuit is shown in Fig. 2-10.

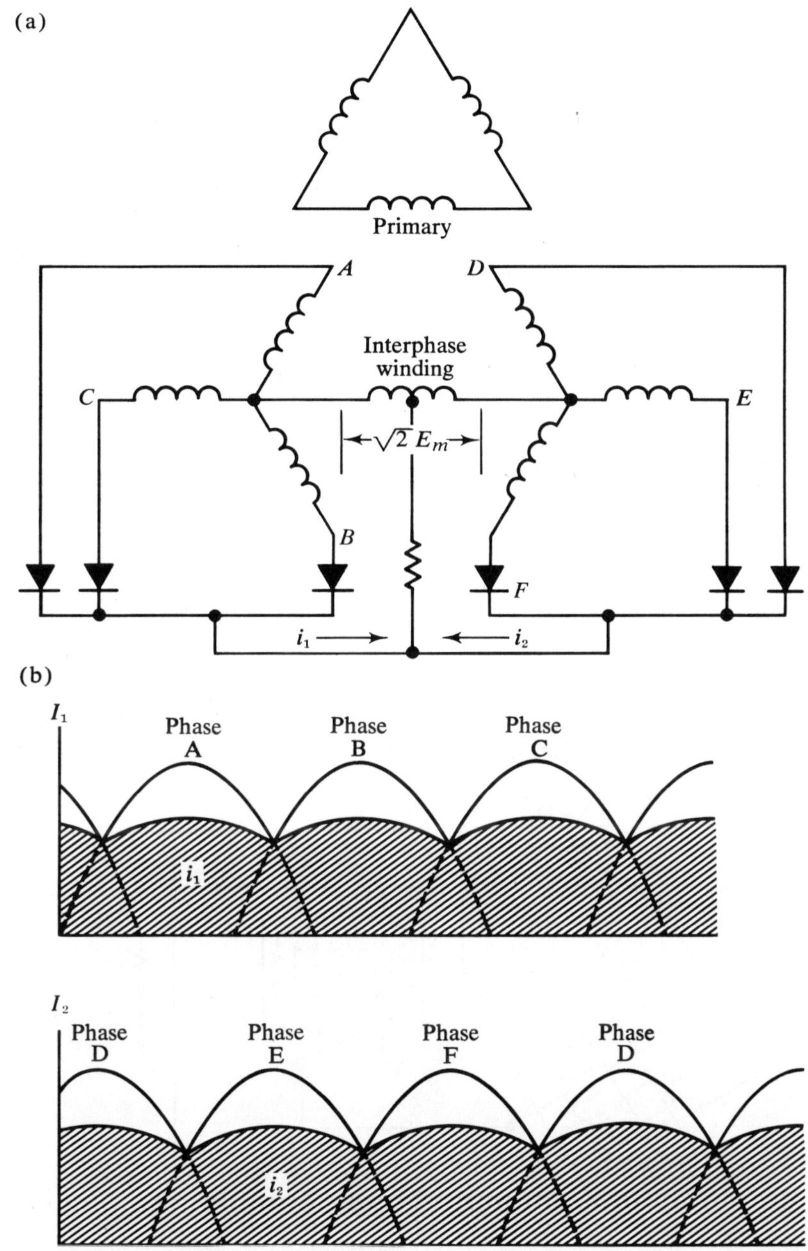

FIG. 2-9. (a) Double-wye rectifier with interphase winding; (b) Current and voltage relationships.

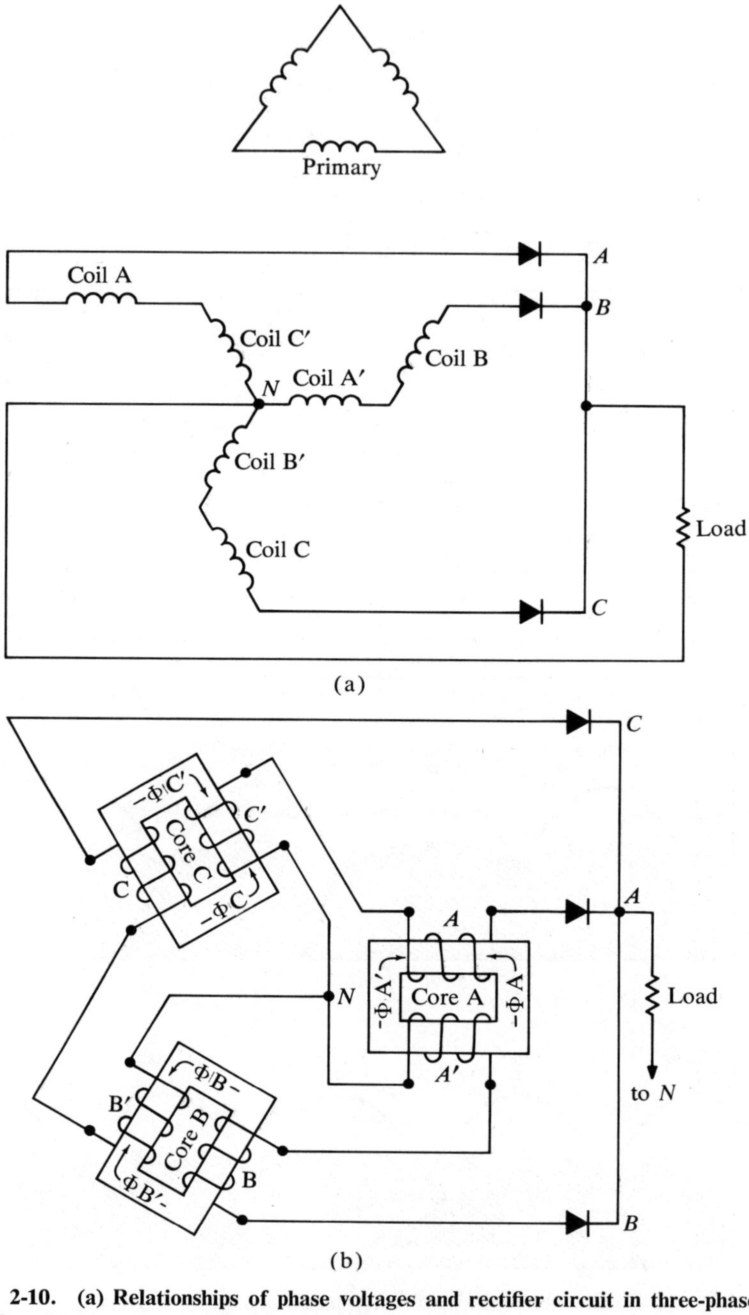

FIG. 2-10. (a) Relationships of phase voltages and rectifier circuit in three-phase zigzag rectifier; (b) Pictorial of winding arrangement on the cores.

By winding one-half of the turns for each secondary on separate cores, d-c saturation of the transformer secondary is avoided. The flux created in the core by each phase is in opposition to the flux created in the same core by the adjoining phase-winding. The d-c components of the flux are effectively cancelled out. Since the voltages from each winding add vectorally to produce the final line-to-neutral voltage, each winding must provide 0.575 of the line-to-neutral voltage.

Problems

1. For the circuit of Fig. 2-1, E_m/phase = 220 v and R_L = 200 ohms. Calculate: (a) the conduction angle for the diodes, (b) the average load voltage, (c) the rms load voltage, (d) magnitude of the ripple voltage, and (e) power in the load.
2. What is the rms current per phase for a transformer supplying the load used for Problem 1?
3. What must the volt-ampere rating for the secondary windings of a transformer be to supply the load used for Problem 1?
4. A single-phase transformer has a secondary voltage rating of 100 v and can deliver 2000 vA. If this transformer is used to supply power to a half-wave rectifier, what value of d-c load resistor is needed to take the rated volt-amperes from the transformer? What amount of average power will this load resistor absorb?
5. A three-phase, half-wave rectifier is delivering 25 A d-c to a 50-ohm resistive load. What is the rms current in each phase winding?
6. What is the magnitude of the a-c ripple current for the rectifier in Problem 5?
7. What is the rms value of the load current in Problem 5?
8. What is the average power in the load for the circuit of Problem 5?
9. A three-phase half-wave rectifier delivers 25 A average current to a load at 500 v average. Calculate the: (1) average current per diode, (2) PIV per diode, (3) volt-ampere rating of the transformer primary, (4) voltage rating of the transformer secondary, (5) rms value of current in the secondary winding, and (6) magnitude of the ripple voltage.
10. A three-phase transformer has a secondary phase voltage of 210 v. What d-c voltage can it deliver to a load if it is used with a three-phase, half-wave rectifier? With a three-phase, full-wave rectifier?
11. A three-phase rectifier bridge is delivering an average d-c voltage of 200 v across a 50-ohm load. Calculate the: (1) maximum voltage across one load, (2) value of the a-c ripple voltage, and (3) average power delivered to the load.

12. The maximum voltage across the load resistor of a three-phase, full-wave bridge is 500 v. What is the magnitude of the a-c ripple voltage?
13. A three-phase, full-wave rectifier delivers 20 A to a load at 175 v (average values). Calculate: the (1) average current/diode, (2) vA rating of the transformer primary, (3) vA rating of the transformer secondary, (4) voltage rating of the secondary winding, (5) rms current in each winding, and (6) magnitude of the ripple voltage.
14. It is desired to deliver 15,000 W to a resistive load. Diodes are available with a 10-A average current rating. What must the voltage and volt-ampere rating of a transformer be if a three-phase, half-wave rectifier circuit is used?
15. Repeat Problem 14 for a three-phase, full-wave bridge.
16. Repeat Problem 14 for a single-phase, half-wave rectifier.
17. Calculate the three-phase, peak-to-peak a-c voltage that is equivalent to a half-wave rectifier output voltage of 10 v rms.
18. Calculate the utility factor of a single-phase transformer that has an rms current of 10 A in its windings and is delivering 1000 W to a 20-ohm load.
19. A three-phase transformer with a phase voltage of 220 v is delivering 2000 W to a resistive load when connected to a three-phase, half-wave rectifier circuit. Calculate the utility factor of this transformer.
20. Repeat Problem 19 for a three-phase, full-wave rectifier with 220 v rms per phase and which is delivering 1500 W to a resistive load.
21. How much power can a 220-v, three-phase transformer deliver to a load if it is connected to a half-wave rectifier and has a winding rated at 10 A rms?
22. Compute the following for a six-phase, half-wave rectifier delivering 100 A d-c to a load: $I_{d\text{-}c}$ per diode, the conduction angle per diode, and the ripple factor.
23. Using the methods described in Sec. 2-2, compute the average output voltage for a six-phase, half-wave rectifier with a phase voltage of 300 v.

Questions

1. Why will each diode conduct for only 120° in a three-phase rectifier circuit? Draw the vector diagram for a three-phase system, and use it to explain your answer.
2. In a three-phase, half-wave rectifier does only one diode conduct at any instant? Why?
3. List the advantages of a three-phase rectifier over a single-phase rectifier.

4. Draw a vector diagram for the voltages in a three-phase system. From this diagram find the peak-inverse voltage across the diodes in a three-phase, half-wave rectifier and in a three-phase, full-wave rectifier.
5. With the vector diagram drawn for Question 4, find the conduction angle for the diodes in a three-phase, half-wave rectifier and in a three-phase, full-wave rectifier.
6. Sketch a six-phase, full-wave rectifier circuit.
7. How does the maximum voltage across the load of a three-phase, half-wave rectifier compare to the maximum voltage across the load of a three-phase, full-wave rectifier? Assume equal phase voltages in each case.
8. Compute the average voltage in a full-wave, single-phase rectifier using Eq. 2-3. Compare the result with the average value of voltage calculated by the method shown in Chapter 1.
9. Referring to Eq. 2-11, explain why the terms I_m and $\sin \omega t$ must be squared to find I_{rms}.
10. Why must the volt-ampere rating of a transformer supplying a half-wave rectifier load be greater than the volt-amperes obtained across the load?
11. Why is a transformer used in a rectifier circuit rated in volt-amperes rather than in watts?
12. What are the ripple frequencies of the a-c voltage in the outputs of a three-phase, half-wave and of a three-phase, full-wave rectifier? Explain, using Figs. 2-1 and 2-3.
13. Referring to Figs. 2-1 and 2-3, indicate the points at which current switches from one diode to another.
14. Why is the ripple factor of a three-phase, half-wave bridge higher than that of a three-phase, full-wave bridge? Explain in terms of Figs. 2-1 and 2-3.
15. Why is the conduction angle for the diodes in a three-phase, half-wave bridge equal to the conduction angle for the diodes in a three-phase, full-wave bridge?
16. Explain why the rms current in each winding of a three-phase transformer is less than the rms value of the load current, even though the maximum values of winding and load currents are equal.
17. Define in your own words "transformer utilization factor."
18. Why are the primary and secondary utilization factors of a transformer different?
19. What is the utilization factor of a single-phase transformer supplying power to a full-wave rectifier circuit? Why?
20. Why will an inductive load improve the utilization factor of a transformer supplying power to a half-wave rectifier circuit?

Magnetic Amplifiers

3

The magnetic amplifier (magamp) is increasingly used in present-day electrical and electronic devices because of its extreme reliability. Recent developments in the design and manufacture of iron cores and diodes have helped to force the magnetic amplifier into prominence as a dependable static device for supplying controlled amounts of power.

In operation, small amounts of power applied to the input of the magamp are used to control much larger load powers. The magnetic amplifier is today one of the most dependable means of power control.

3-1 MAGNETIC-AMPLIFIER TERMINOLOGY

Figure 3-1 shows a diagram of a basic magnetic amplifier with the components labeled.

The load winding, wound around an iron core, presents a large inductive reactance to the load current. The magnitude of this reactance can be varied by passage of a d-c current through the control winding. This winding usually consists of a very large number of turns, permitting small amounts of d-c current to drive the core into a saturated state. When the core is saturated, the inductive reactance, in series with the load, approaches zero.

The rectifier is placed in series with the load in order to reduce the amount of power needed in the control winding to saturate the core. This procedure will be discussed in detail in the following paragraphs.

The iron-core material should have a hysteresis loop that resembles the ideal loop shown in Fig. 3-1(b). The ideal loop can only be produced by a material that requires no control power to be driven into saturation. Thus the ideal loop is impossible to achieve in practice. A practical iron-core hysteresis loop may, however, approximate that shown in Fig. 3-1(c). Notice

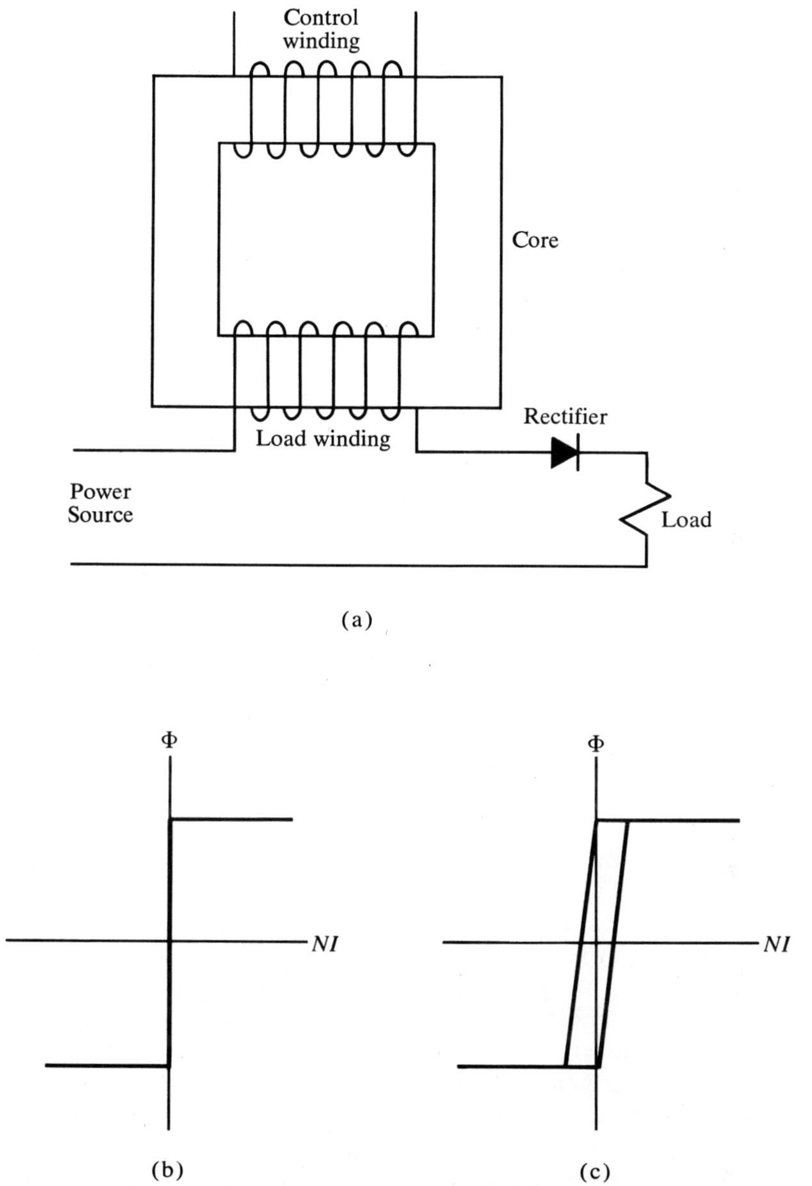

FIG. 3-1. (a) Magnetic amplifier; (b) Ideal hysteresis curve for magnetic-amplifier core material; (c) Practical-amplifier cores can approach this "box-shaped" hysteresis loop.

that the loop of Fig. 3-1(c) has a very sharply defined saturation region and a narrow hysteresis curve.

The saturable reactor, the forerunner of the magnetic amplifier, is similar to the magnetic amplifier of Fig. 3-1 except for absence of a rectifier. More control power is needed to saturate the core, and the operation of this device is not as efficient as that of the magnetic amplifier.

3-2 REVIEW OF INDUCTANCE

Inductance has been described as the property of an electrical circuit that opposes any change in current through that circuit. The inductance of an iron-core coil can be expressed as

$$L = \frac{1.257AN^2\mu}{l_m \times 10^8} \tag{3-1}$$

where L = inductance in henrys
N = number of turns
μ = permeability of the core material in mks units
A = cross-sectional area of the core in square centimeters
1.257 = a constant
10^8 = a constant
l_m = length of the magnetic circuit in centimeters

The factors A, l_m, and N are constants that depend upon the physical construction of the core; μ depends upon the type of core material and the flux density in the core. The inductance of the coil can, therefore, be controlled by varying the permeability of the iron core. The BH curve for an electromagnetic coil with iron core is shown in Fig. 3-2(a).

Assuming that an alternating input current starts from zero in a positive direction, the core flux will start at zero and rise to point A at the peak of the cycle. At point A the permeability of the core is very low, and any additional increases in magnetizing force will not greatly increase the flux density. At this point the inductance of the coil is low. When the input current drops to point b, after one-half cycle, the flux density in the core will not return to zero but will remain at point B on the hysteresis curve. The flux density would remain at this level (B) if the magnetizing current were removed from the core at this point. Such remaining flux is called *residual flux* or *residual magnetism*, and the ability of a core to retain magnetism after the magnetizing force has been removed is called *retentivity*.

When the current has reversed and reaches the value at point c, the flux

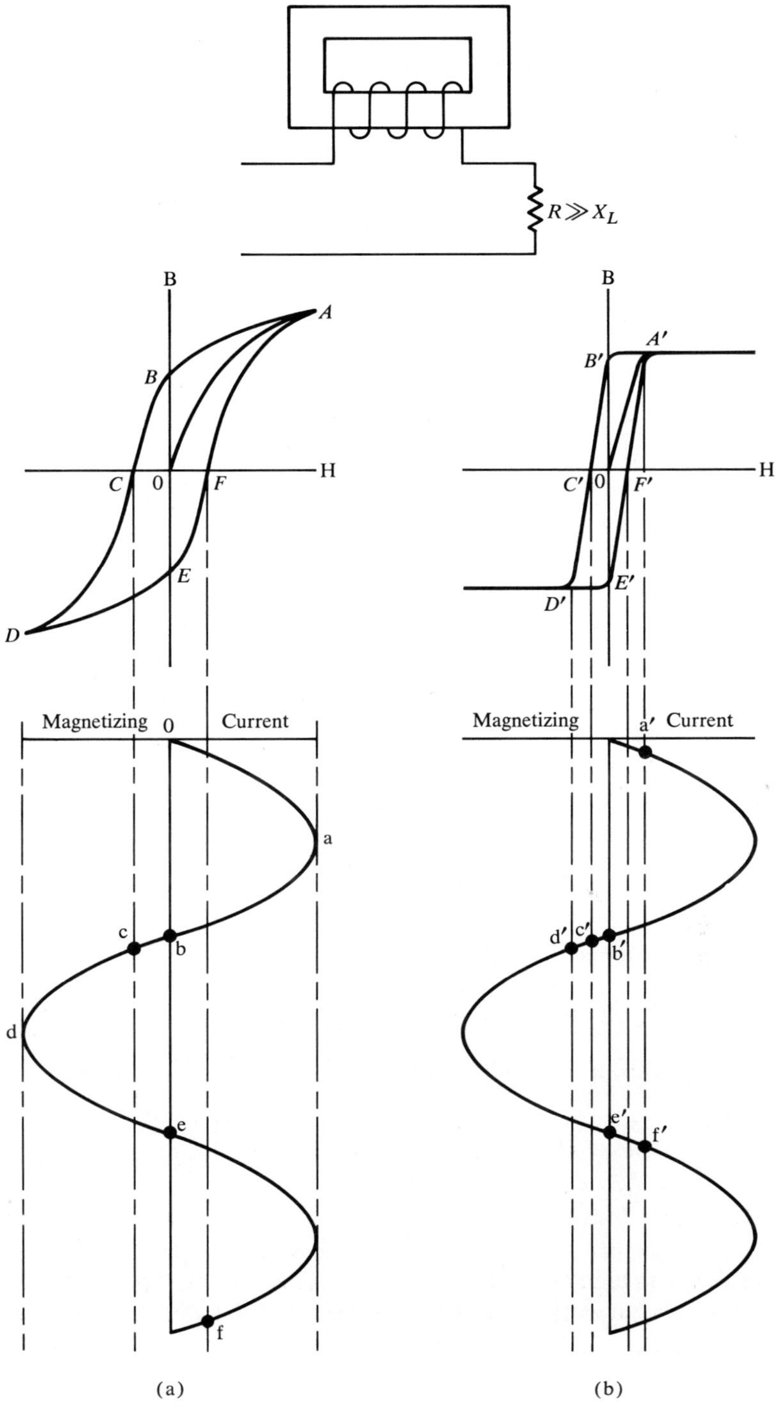

FIG. 3-2. (a) Hysteresis loop for transformer core material; (b) "Box-shaped" hysteresis loop for material used in magnetic amplifiers.

density in the core is reduced to zero. The magnetizing force needed to reduce the field to zero is called *coercive force*.

When the input current reaches point *d*, the core is again saturated but in the negative direction. The magnetizing force now moves toward zero from its maximum negative value at point *d*. At point *E* the core is again left with a residual flux but with zero magnetizing force.

If the core has the idealized, narrow box-shaped hysteresis loop of Fig. 3-2(b), a small magnetizing current will saturate the core, and the inductance will be low for almost the full cycle. When the current rises to point a', the core will become saturated and will remain so until the magnetizing force has actually become negative, as at point c'. From points B' to D' the inductance of the core will be high because of the rapid change in flux. At point D' the core will saturate, and again the inductance will be low. It should be observed that after the core is saturated, the inductance will remain low until the flux is reduced by an opposing magnetizing force.

3-3 THE IRON-CORE COIL

A study of magnetic amplifiers should begin with an investigation of a simple iron-core coil, which is the fundamental component of both the saturable reactor and the magnetic amplifier.

In the circuit shown in Fig. 3-3, the current to the load will depend upon the value of the load resistor and upon the inductance of the coil. The basic formula for the inductance of an iron-core coil is

$$L = \frac{1.257 A N^2 \mu}{l_m \times 10^{+8}} \qquad \text{from Eq. 3-1}$$

The induced voltage in the coil which opposes current flow is

$$e = N \frac{d\phi}{dt} \tag{3-2}$$

From Fig. 3-3(b) μ is calculated as

$$\mu = \frac{\Delta \mathrm{B}}{\Delta \mathrm{H}} \tag{3-3}$$

or

$$\mu = \frac{\Delta \mathrm{B}}{\Delta_{\mathrm{I}} NI} \tag{3-4}$$

where the change in ampere turns NI is a function of the peak-to-peak value of the a-c current in the coil. It can be seen that the μ of the core and, therefore, the inductance will be constant as long as the a-c current in the core

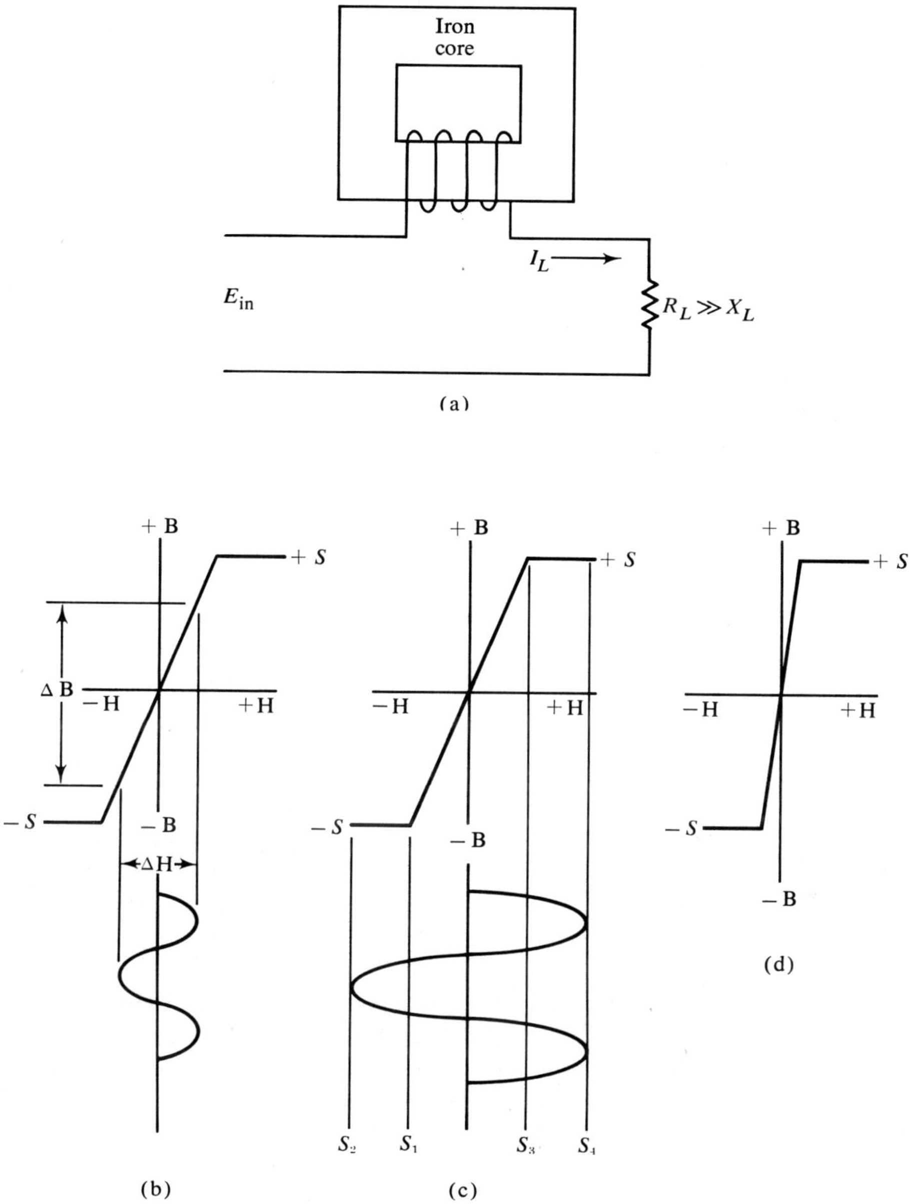

FIG. 3-3. (a) Iron-core inductor with load; (b) Low-load current, nonsaturating; (c) High-load current-saturating; (d) Steep hysteresis loop, low-load current required for saturation.

does not drive the flux density beyond the saturation points $+S$ and $-S$. Should the a-c current saturate the core during any portion of the cycle, the *average* value of μ will be decreased and the *average* inductance of the core lowered.

From Eq. 3-2 it is seen that the induced voltage which opposes current is a function of the rate of flux change in the core. If the core does not saturate, the rate of flux change is a function of the change of input current only.

If the coil current saturates the core, as shown in Fig. 3-3(c), the induced voltage is zero in that period of the cycle between points $S1$ & $S2$ and also between $S3$ & $S4$, because between each pair of points the flux change is zero. For any portion of the cycle, during which the rate of change in flux is zero, the inductance of the core is also zero. The flux change in the core and the inductance of the coil remains at zero until flux is reduced below the saturation value.

If the slope of the BH curve is steep, as in Fig. 3-3(d), very little current will be necessary to produce saturation, and the reactance of the coil can easily be driven to zero.

To analyze the behavior of the current in the iron-core coil, refer to Fig. 3-4. When the core saturates, the reactance is zero, and the current will quickly rise to a value equal to approximately E_{in}/R_L.

Starting at $t = 0$ and $I = 0$, the current is opposed by the inductance of the reactor. At α_1 the current is high enough to drive the core into saturation, and the reactance is reduced to zero. At this point (α_1) the current rises rapidly to its maximum value and then follows the input voltage waveform.

As the input voltage decreases the current follows the voltage wave to point B. At that point the core will no longer be saturated, and the inductance of the coil will increase. The energy stored in the magnetic field will induce a voltage in the coil. This voltage will cause the load current to remain positive until the negative value of the input is larger than the induced positive voltage.

Refer again to Fig. 3-4(b). The angle ϕ_f is called the *firing angle*. This angle is a measure of the angular distance through which the input voltage vector must rotate before the core is driven to saturation. The firing angle varies according to magnitude of applied voltage, size of core, and value of load resistance. The equation for the firing angle ϕ_f is derived as follows:

In the iron-core coil

$$e = N\frac{d\phi}{dt}$$

$$\Phi = \frac{1}{N}\int e\, dt \qquad \textbf{(3-5)}$$

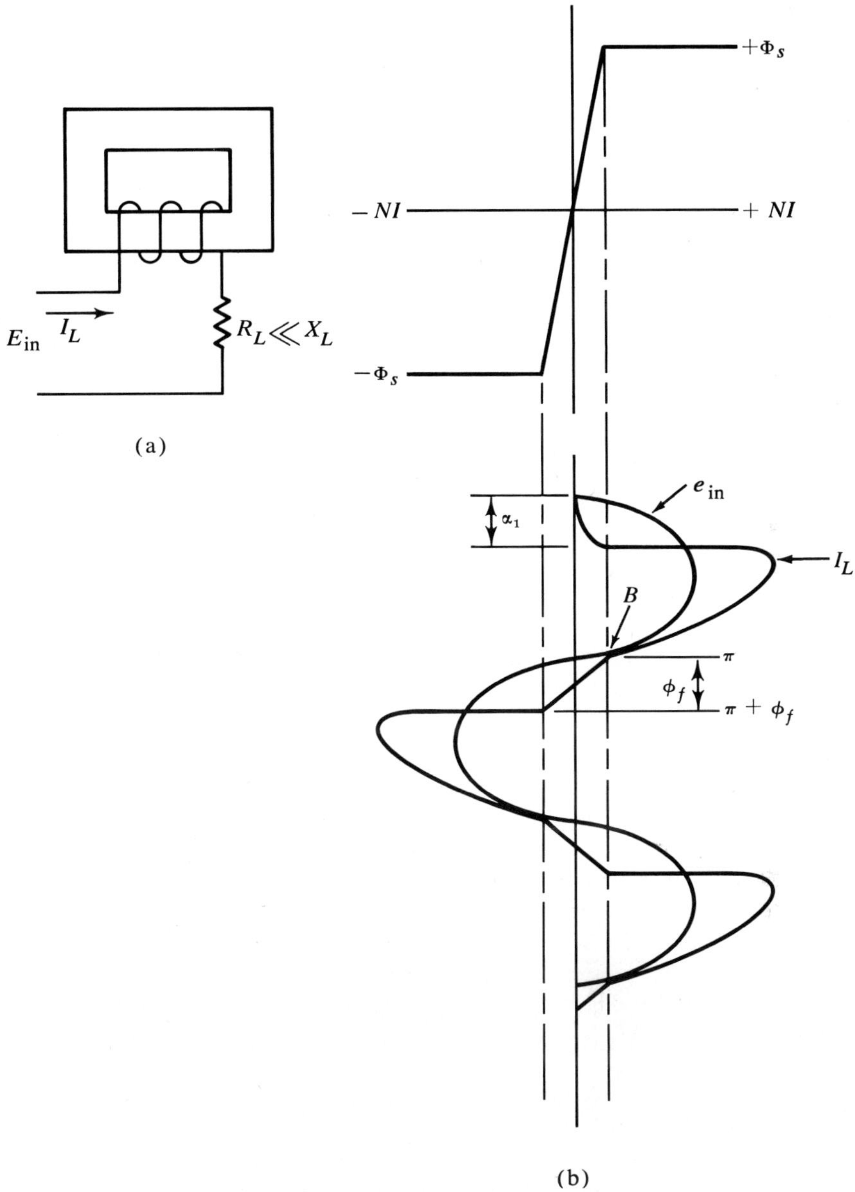

FIG. 3-4. (a) Value of X_L determines magnitude of I_L; (b) Current and voltage waveforms when $R_L \ll X_L$.

For a sinusoidal input voltage

$$\Phi\ (\text{webers}) = \frac{1}{N}\int E_{\mathrm{m}} \sin \omega t(dt) \tag{3-6}$$

or

$$\Phi = -\frac{E_{\mathrm{m}}}{\omega N}\cos \omega t + K_1 \tag{3-7}$$

This is the instantaneous magnitude of flux in the core.

Setting the initial conditions at $t = 0$ and $\Phi = 0$,

$$K_1 = \frac{E_{\mathrm{m}}}{\omega N} \tag{3-8}$$

Rewriting Eq. 3-7

$$\Phi = -\frac{E_{\mathrm{m}}}{\omega N}\cos \omega t + \frac{E_{\mathrm{m}}}{\omega N} \tag{3-9}$$

which reduces to

$$\Phi = \frac{E_{\mathrm{m}}}{\omega N}(1 - \cos \omega t) \tag{3-10}$$

At $\omega t = \pi$ the flux in the core is equal to the saturation flux Φ_s. Between $\pi < \omega t < \pi + \phi_f$, the instantaneous flux in the core is,

$$\Phi = \Phi_s - \frac{E_{\mathrm{m}}}{\omega N}(1 - \cos \omega t) \tag{3-11}$$

and at $\omega t = \phi_f$

$$-\Phi_s = \Phi_s - \frac{E_{\mathrm{m}}}{\omega N}(1 - \cos \phi_f)$$

Solving for ϕ_f,

$$\phi_f = \cos^{-1}\left(1 - \frac{2\Phi_s \omega N}{E_{\mathrm{m}}}\right) \tag{3-12}$$

Equation 3-12 states that the firing angle measured from the beginning of each half-cycle will increase if the flux required for saturation is increased. The firing angle will be reduced as the input voltage is increased.

3-4 THE SATURABLE-CORE REACTOR

The saturable-core reactor, or saturable reactor, is a device used to control large amounts of a-c power by means of small quantities of d-c input power. The saturable reactor has a distinct advantage compared to control rheostats in that its power losses are relatively small. In the saturable reactor, the inductive reactance is varied. Since true power is not consumed in an inductance, none of the I^2R losses associated with rheostat controls occur in the case of saturable reactors. This reactor also has the advantage, mentioned above, of being controllable by a small amount of d-c power. The d-c control power may be varied at a location remote from the load or may be controlled automatically by an error-signal from the load. A saturable reactor with d-c-control winding and a-c load winding is shown in Fig. 3-5.

The behavior of current in the load circuit has been discussed in the preceding paragraph; a direct-current-control winding will now be added to the iron core. This winding, consisting of many turns of fine wire, is designed to enable a current of low magnitude to drive the core to saturation. This small direct current can control large amounts of alternating-current power in the load circuit if the reactor is properly designed. The current in the load winding will induce an a-c voltage in the control circuit. The blocking choke is, therefore, placed in the control circuit to prevent the circulation of induced a-c currents.

When a d-c current flows in the control winding, the control ampere-turns NI_c, as shown in Fig. 3-5(b), shifts the operating point of the iron core from the zero intersect of the magnetization curve to point Φ_0. Any flux established by the load current will then add or subtract from the flux established by the control winding, as described below.

As a result of biasing the core at point Φ_o, the load current waveform will have the appearance shown in Fig. 3-5(b). As the load current rises toward its maximum positive value, the core will saturate when the ampere-turns, established by the load current, and the control-current ampere-turns add to produce the necessary magnetizing force. At the saturation point the core will *fire*, and the current will rise to its maximum value. When the load current reverses, the load and control fluxes will oppose each other and prevent the core from saturating. The load current will be near zero for this half-cycle.

Comparison of the current waveform of Fig. 3-5(b) with that of Fig. 3-5(c), shows that the value of the average load current has been changed by the change of current in the d-c control winding. The core, shown in Fig. 3-5(c), is operated about the zero intersect of the magnetization curve, and the input current is not of large enough magnitude to saturate the core. Since the inductance of the core is high for the full cycle, the load current is small. When a direct current is applied to the core and its bias point is shifted to point Φ_o of Fig. 3-5(b), the core is saturated for almost a full half-cycle, and the average load current is increased.

The current required to change the flux density in the core from point Φ_o to saturation point Φ_s determines the firing angle for the input voltage wave. The angle ϕ_f will be a function of the change in flux.

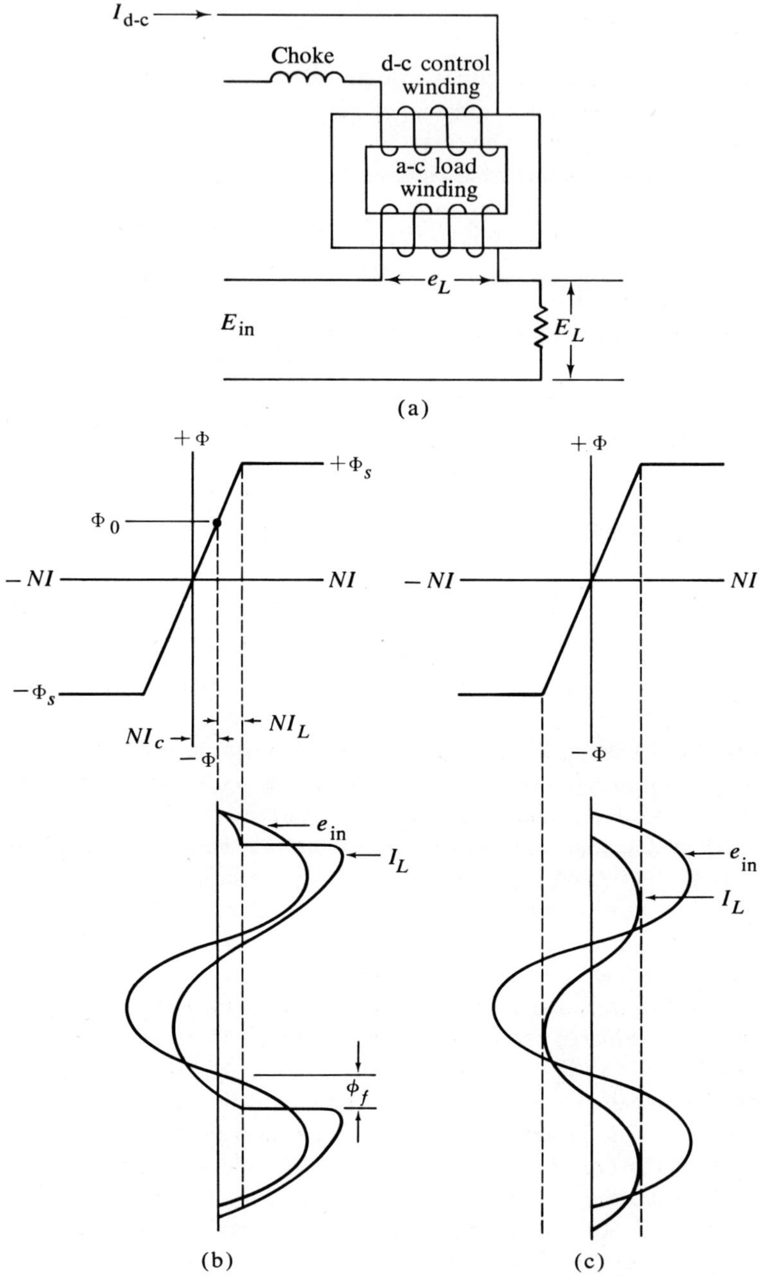

FIG. 3-5. (a) Saturable reactor with d-c control winding; (b) Voltage and current waveforms when $I_{d\text{-}c}$ has shifted operating point to Φ_o; (c) Operation of saturable reactor about the zero intersect of the magnetization curve.

From Fig. 3-5b

$$\Delta\Phi = \Phi_s - \Phi_o$$

From Eq. 3-6,

$$\Delta\Phi = \frac{E_m}{N_L}\int_0^{\phi f} \sin \omega t(dt)$$

Therefore,

$$\Phi_s - \Phi_o = \frac{E_m}{\omega N_L}(1 - \cos \phi f)$$

If the core reaches saturation at $\omega t = \pi/2$, then, from Eq. 3-9

$$\Phi_s = E_m/\omega N_L \quad \text{and} \quad \Phi_s - \Phi_o = \Phi_s(1 - \cos \phi_f)$$

Solving for ϕ_f,

$$\phi_f = \cos^{-1}\frac{\Phi_o}{\Phi_s} \tag{3-13}$$

Equation 3-13 is based on the requirement that the input voltage can barely drive the core to saturation at E_m or

$$\Phi_s = \frac{E_m}{\omega N_L} \tag{3-14}$$

Example 3-1 If the flux density required to saturate the core of the magnetic amplifier shown in Fig. 3-5 is 25,000 lines/in.², at what angle will the reactor fire if the control winding biases the core at the following values of B_o: (a) Zero? (b) 5000 lines/in.²? (c) 15,000 lines/in.²? (d) 23,000 lines/in.²? Assume that the maximum input voltage will drive the core to saturation when $B_o = 0$.

From Eq. 3-13, $\phi_f = \cos^{-1}\frac{B_o}{B_s}$

when $B_o = 0$ and $\phi_f = \cos^{-1} 0$, $\phi_f = 90°$
when $B_o = 5{,}000$ and $\phi_f = \cos^{-1} 0.2$, $\phi_f = 78.5°$
when $B_o = 15{,}000$ and $\phi_f = \cos^{-1} 0.6$, $\phi_f = 53°$
when $B_o = 23{,}000$ and $\phi_f = \cos^{-1} 0.92$, $\phi_f = 23°$

3-5 TWO COILS ON A SINGLE CORE

When two coils (such as the load- and control-windings of a saturable reactor) are wound on a common iron core, an a-c current flowing in one winding will induce an a-c voltage in the other. Such transformer action between the coils reduces the efficiency of the saturable reactor because the load current will induce a large a-c current in the d-c control circuit. During one-half of the cycle such induced current will oppose the d-c-control current, and extra power will be required from the d-c source to overcome it. Figure 3-6 shows an iron-core reactor wound with two coils.

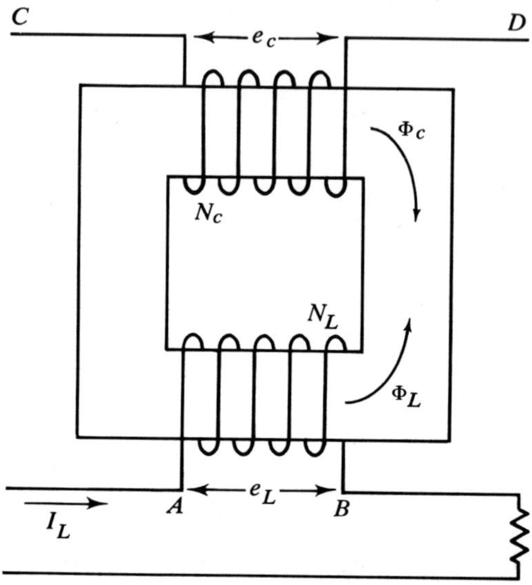

FIG. 3-6. Iron core with load and control windings.

The two-coil winding can be analyzed with the methods applied to an iron-core transformer as long as the core is not saturated. From general transformer theory we can state:

$$\frac{I_c}{I_L} = \frac{N_L}{N_c}$$

where

$$\begin{aligned} I_L &= \text{load current} \\ N_L &= \text{load winding turns} \\ N_c &= \text{control winding turns} \\ I_c &= \text{current induced in } N_c \text{ by } I_L \end{aligned}$$

The induced current I_c will act in every instance to reduce the flux change that produced it. This action will in turn increase the control current needed to drive the core to saturation.

The current I_c flows through an impedance Z_c of the control circuit. Therefore,

$$I_c = \frac{e_c}{Z_c} \tag{3-15}$$

By transformer action,

$$\frac{e_L}{e_c} = \frac{N_L}{N_c}$$

and

$$e_c = e_L \frac{N_c}{N_L} \tag{3-16}$$

Combining Eqs. 3-15 and 3-16

$$I_c = \frac{N_c}{N_L} \times \frac{e_L}{Z_c} \tag{3-17}$$

From transformer action

$$\frac{I_L}{I_c} = \frac{N_c}{N_L} \tag{3-18}$$

and

$$I_L = I_c \frac{N_c}{N_L} \tag{3-19}$$

Combining Eqs. 3-17 and 3-19

$$I_L = \frac{N_c}{N_L} \times \frac{e_L}{Z_c} \times \frac{N_c}{N_L} \tag{3-20}$$

or

$$I_L = \left(\frac{N_c}{N_L}\right)^2 \times \frac{e_L}{Z_c} \tag{3-21}$$

The a-c input impedance for the load winding becomes

$$\frac{e_L}{I_L} = Z_c \left(\frac{N_L}{N_c}\right)^2 \tag{3-22}$$

The impedance of the load circuit is reflected to the control circuit and the control-circuit impedance is reflected to the load circuit.

Example 3-2 The reactor core shown in Fig. 3-6 consists of a load winding having 200 turns and a resistance of 2 ohms. The control winding has 2000

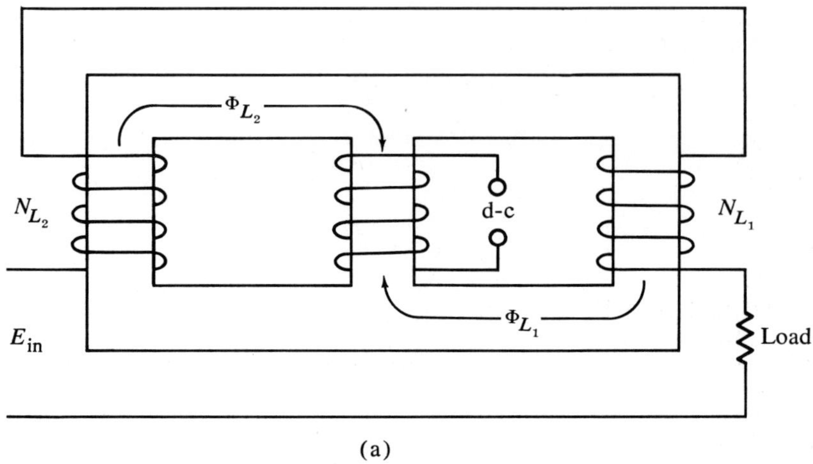

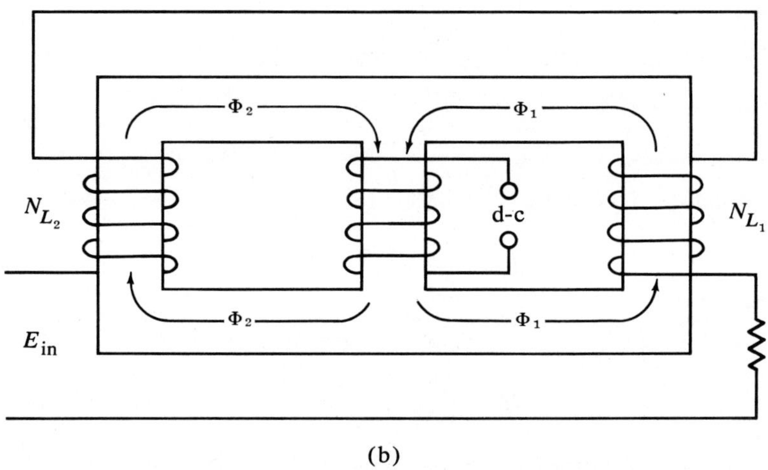

FIG. 3-7. (a) Nonpolarized core; (b) Polarized core.

turns and a resistance of 30 ohms. What is the impedance at the terminals A and B of the load winding when current first starts to flow in the load? From Eq. 3-22

$$Z_{AB} = 30\left(\frac{200}{2000}\right)^2 = 0.3 \text{ ohms}$$

If the control winding has a 30-ohm resistance, what will the losses in this winding be when $I_L = 20$ A and the core is not saturated?

From Eq. 3-18

$$I_c = I_L \frac{N_L}{N_c}$$

$$I_c = 20\left(\frac{200}{2000}\right)$$

$$I_c = 2 \text{ A}$$

$$\text{w} = I^2R$$

$$\text{w} = 4(30)$$

$$\text{w} = 120 \text{ W}$$

One method of reducing the I^2R losses caused by induced currents in the control winding is to use the three-legged core shown in Fig. 3-7(a).

The load windings are wound around the two outside legs of the core in such a manner that the lines of flux produced in each are opposed in the center leg. The control winding is wound on the center leg and is not cut by the flux linkages created by the change in load current. When the load

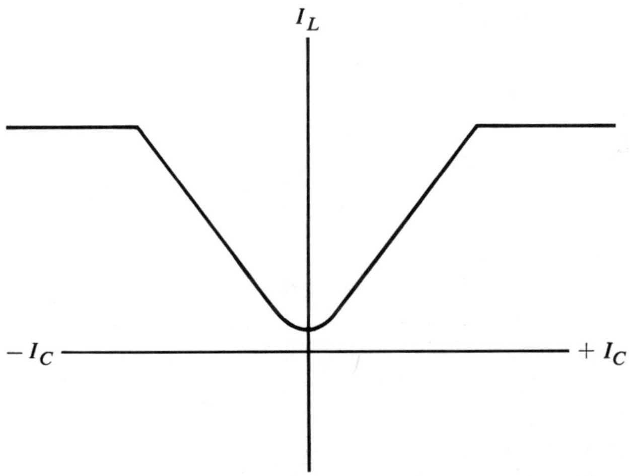

FIG. 3-8. Control characteristics, nonpolarized core.

current reverses, the lines of flux produced in the outside legs are still in opposition at the center leg.

When a direct current is used in the control winding to saturate the core, the flux produced by the load current aids the control flux for one half-cycle and opposes it for the other half-cycle. The operation will be the same for any direction of current in the control winding, hence, the name *nonpolarized* reactor. The dynamic characteristic curve for a nonpolarized reactor is given in Fig. 3-8.

In some cases it is necessary to use the *polarized core*, shown in Fig. 3-7(b). This type of core is wound in a way to permit the lines of flux produced in the two outside legs to be added in the center leg. The lines of flux produced by the load-windings may aid the control flux or oppose it, depending upon the direction of the load current, hence, the name polarized core.

3-6 THE EFFECT OF A DIODE ON THE IRON-CORE REACTOR

When a diode is added in series with the load coil of a saturable-reactor circuit, as shown in Fig. 3-9, the effect is to cancel out the inductance of the core after the first half-cycle of current.

The rectifier element allows current through the core in one direction only. The reactor fires at an angle ωt, equal to $\cos^{-1}[1 - (2\Phi_s\omega N/E_m)]$, as given previously by Eq. 3-12. The core is saturated at $\omega t = \phi_f$, as shown in Fig. 3-9(b). The current in the load then follows the sinusoidal input. When the input voltage reverses, the rectifier prevents the current from reversing, and the load current is zero. The iron core remains in saturation because there is no reverse current to reduce the residual flux. When the voltage input again moves in the positive direction, no opposition to current flow is caused by the inductance of the core since the core is already at saturation. Therefore, unless the flux in the core is *reset* to zero, the circuit will function as a normal half-wave rectifier circuit.

When a direct current is supplied to the control winding, the magnetizing force produced by the control current can be used to *reset* the core after each cycle. This operation is shown in Fig. 3-9(c). The control-ampere turns NI_c are impressed on the core in a direction that opposes the load-ampere turns. When the input voltage becomes positive, the core fires as before; however, the firing angle ϕ_f is increased as a direct result of the biasing effect of the control-ampere turns. When the input voltage returns to zero, the control-ampere turns reverse the net magnetizing force on the core and reduce the residual flux to zero. No current flows in the load winding when the a-c input

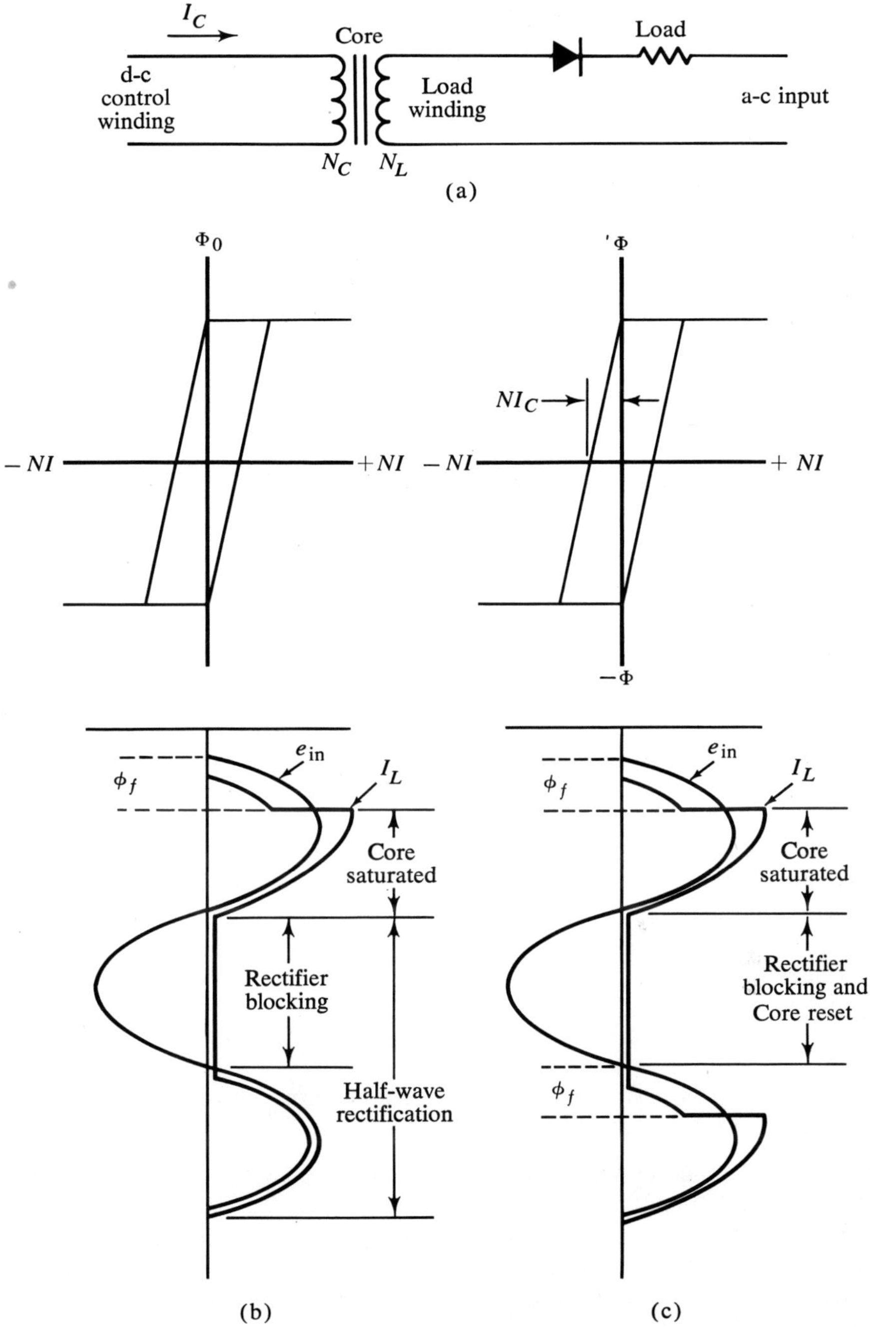

FIG. 3-9. (a) Schematic diagram of basic magnetic-amplifier; (b) Waveforms with blocking rectifier; (c) Waveforms with blocking rectifier and control current *reset*.

is negative because of the blocking action of the diode. When the input voltage becomes positive in the next cycle, the core will again fire and reproduce the current waveform of the first cycle. It should be obvious to the student that the firing angle and, therefore, the effective load current can be controlled by the amount of reverse bias impressed on the core by the control winding.

3-7 THE BASIC MAGNETIC-AMPLIFIER

The basic magnetic-amplifier, consisting of two saturable reactors connected in series, is shown in Fig. 3-10.

Both *load windings* N_{L_1} and N_{L_2} are connected in series and the *control windings* N_{C_1} and N_{C_2} are also connected in series. The polarities of the windings are such that while current flows in the load windings N_{L_1} and N_{L_2}, the a-c voltages induced in windings N_{C_1} and N_{C_2} oppose each other, and the induced a-c current in the control circuit is zero. These conditions occur only when core 1 and core 2 are operating in a nonsaturated portion of the magnetization curve.

A basic design-requirement for proper operation is that, at zero-control current, the a-c input current must produce a flux density in the reactor cores, and that such density must be almost equal to but below the saturation level of the cores. Since each core is unsaturated, the reactance of each is high, and very little current flows into the load. Note that when a direct current is supplied to the control windings in a direction that causes positive bias of core 1, core 2 will be biased negative by the same amount. The bias points, $+a$ and $-a$, are shown in parts (b) and (c) of Fig. 3-10.

When the input voltage becomes positive, core 1 will saturate when the flux reaches Φ_s. The firing angle at which the flux equals Φ_s can be expressed as follows:

$$\phi_f = \cos^{-1}\left(\frac{\Phi_o}{\Phi_s}\right) \tag{3-23}$$

where Φ_o = flux produced by the d-c-control current

Φ_s = saturation flux

When core 1 becomes saturated, the voltage induced into control winding N_{C_1} becomes zero, and an a-c current flows in the control circuit. This current will have the same waveshape as the load current I_L. The induced current $I_{\text{a-c}}$ moves in a direction that aids the d-c-control current of core 2 and produces a magnetizing force $N_{C_2}I_{\text{a-c}}$ that is in opposition to the load current magnetizing force $I_L N_{L_2}$ of core 2. Since the flux produced by the a-c current in the control winding opposes the flux produced by the current in the load winding, the net change in flux in core 2 is zero, and its reactance is zero.

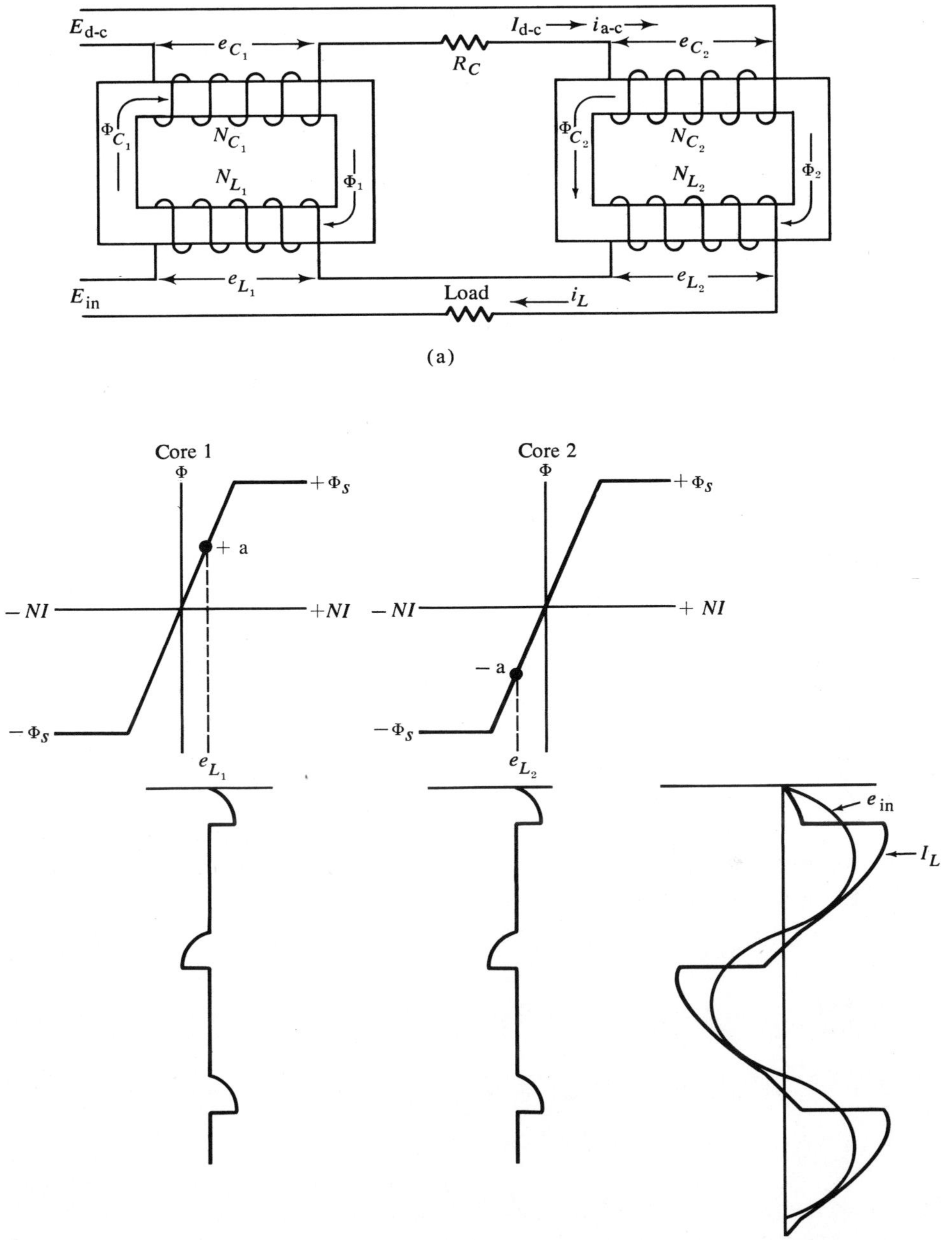

FIG. 3-10. (a) Basic magnetic-amplifier; (b) Flux and voltage relationships in core 1; (c) Flux and voltage relationships in core 2; (d) Input-voltage- and load-current waveforms.

The source voltage will appear across the load, and the load current will be approximately equal to $E_{a\text{-}c}/R_{load}$. Both the resistance of the load windings and the reflected resistance of the control windings have an effect on the load current. Note that since the value of the d-c-control current can cause variation in the angle at which the cores fire, control of the effective value of the load current is easily obtained. It is important to observe that the control current is composed of two components: $I_{d\text{-}c}$ which is supplied by the d-c source and $I_{a\text{-}c}$ which is induced current. $I_{a\text{-}c}$ is usually much larger than $I_{d\text{-}c}$.

3-8 POWER- AND CURRENT-AMPLIFICATION

The power gain W of a magnetic amplifier is defined as

$$W = \frac{\text{useful load power}}{\text{control circuit power}} = \frac{P_L}{P_c} \tag{3-24}$$

where

$$P_L = I_L{}^2 R_L$$

and

$$P_c = I_c{}^2 R_c$$

By substitution,

$$W = \left(\frac{I_L}{I_c}\right)^2 \frac{R_L}{R_c} \tag{3-25}$$

I_L and I_c being effective values.

Since control and load windings are wound on the same core, the transformer action between the windings relates the control and load currents by the well-known relationship:

$$\frac{N_C}{N_L} = \frac{I_L}{I_C}$$

The current gain, K_c, is thereby a function of the turns ratio of N_C and N_L.

$$K_c = \frac{N_c}{N_L} \tag{3-26}$$

The power gain may now be expressed as

$$W \cong \left(\frac{N_c}{N_L}\right)^2 \frac{R_L}{R_c} \tag{3-27}$$

The approximately equal sign is used because this relationship will be affected by both the waveform of the current and the operating conditions of the core.

Example 3-3 A magnetic amplifier consists of a load winding of 800 turns and a control winding of 6400 turns. Assuming the resistance of the control winding to be 20 ohms and the load resistance 10 ohms, calculate: (a) the current gain of the amplifier and (b) the power gain of the amplifier.

1. From Eq. 3-26

$$K_c = \frac{N_c}{N_L}$$

$$K_c = \frac{6400}{800} = 8$$

2. From Eq. 3-27

$$W \cong \left(\frac{N_c}{N_L}\right)^2 \frac{R_L}{R_c}$$

$$W \cong \left(\frac{6400}{800}\right)^2 \frac{10}{20}$$

$$W \cong 32$$

3-9 TIME-CONSTANT CHARACTERISTICS

The time constant of a magnetic amplifier is important when the amplifier is used in applications specifying feedback control and equally sophisticated circuits. We know that for any inductive circuit the time constant can be expressed as follows: $T = L/R$. The same equation is true of the magnetic amplifier when L and R are properly defined. The general expression for the inductance of the control winding can be expressed as

$$L_c = \frac{\Delta(N_c\phi)}{\Delta I_c} \tag{3-28}$$

In magnetic amplifiers, flux varies from zero to saturation continuously; therefore, it is necessary to determine average flux and its rate of change. As shown in Eq. 3-10, the saturation flux is related to the input frequency as follows (assuming input current barely saturates the core at $\pi/2$).

$$\Phi_s = \frac{E_{\mathrm{m}}}{\omega N_L} \tag{3-29}$$

For half-cycle conduction the flux varies from zero to $E_{\mathrm{m}}/\omega N_L$.

The average flux is therefore

$$\Phi_{\mathrm{av}} = \frac{E_{\mathrm{m}}}{2\omega N_L} \tag{3-30}$$

If

$$I_c = \frac{N_L}{N_c} I_L \tag{3-31}$$

and the average current I_L is expressed as $2E_m/\pi R_L$, then the average current per cycle is

$$I_c = \frac{N_L}{N_c} \times \frac{2E_m}{R_L \pi} \tag{3-32}$$

Combining Eqs. 3-28, 3-30, and 3-32, and bearing in mind that there are two cores in series, the total inductance becomes

$$L_c = \frac{2\Delta(N\phi)}{\Delta I_c} = \frac{2N_c\left(\dfrac{E_m}{4\pi f N_L}\right)}{\dfrac{N_L}{N_c} \times \dfrac{2E_m}{\pi R_L}}$$

or

$$L_c = \frac{R_L}{4f}\left(\frac{N_c}{N_L}\right)^2 \tag{3-33}$$

Equation 3-33 describes the inductance of the core looking into the control winding to which all control signals are applied. The time constant L/R_c becomes

$$T_c \cong \frac{R_L}{R_c} \times \left(\frac{N_c}{N_L}\right)^2 \times \frac{1}{4f} \tag{3-34}$$

The first two terms of Eq. 3-34 can be recognized as the power gain of the magnetic amplifier.

It should be noted that the time constant is a function of power gain and input frequency. This relationship unfortunately limits the use of a magnetic amplifier. Whenever the power amplification is increased, the time constant increases and the frequency response is lowered.

Example 3-4 A magnetic amplifier consists of a control winding having 1,000 turns and a 5-ohm resistance and of a load winding having 20 turns and a 0.2-ohm resistance. The device is operating from a 60-Hz line into a 0.5-ohm load. Calculate the:

1. inductance of the control winding
2. current gain of the amplifier
3. power gain of the amplifier
4. time constant of the amplifier

1. From Eq. 3-33

$$L_c = \frac{R_L}{4f}\left(\frac{N_c}{N_L}\right)^2$$

$$L_c = \frac{5}{4 \times 60}\left(\frac{1000}{20}\right)^2$$

$$L_c = 52 \text{ H}$$

2. From Eq. 3-26

$$K_c = \frac{N_c}{N_L}$$

$$K_c = \frac{1000}{20}$$

$$K_c = 50$$

3. From Eq. 3-27

$$W \cong \left(\frac{N_c}{N_L}\right)^2 \frac{R_L}{R_c}$$

$$W \cong \left(\frac{1000}{20}\right)^2 \frac{0.5}{5}$$

$$W \cong 250$$

4. From Eq. 3-34

$$T_c \cong \left(\frac{N_c}{N_L}\right)^2 \times \frac{R_L}{R_c} \times \frac{1}{4f}$$

$$T_c \cong 250(\tfrac{1}{240})$$

$$T_c \cong 1.04 \text{ sec}$$

3-10 MAGNETIC AMPLIFIER WITH EXTERNAL FEEDBACK

The use of external feedback can greatly increase the gain and usefulness of the magnetic amplifier discussed in the preceding section. Positive feedback is generally used because it is desirable to increase the gain of the magnetic amplifier; bandpass characteristics are usually not of primary importance. In Fig. 3-11 a circuit employing positive feedback operation is shown.

This circuit is essentially the same as that of Fig. 3-10, except for the addition in the former of feedback windings. Each winding is wound on its core in a direction that causes production of an a-c voltage in the winding

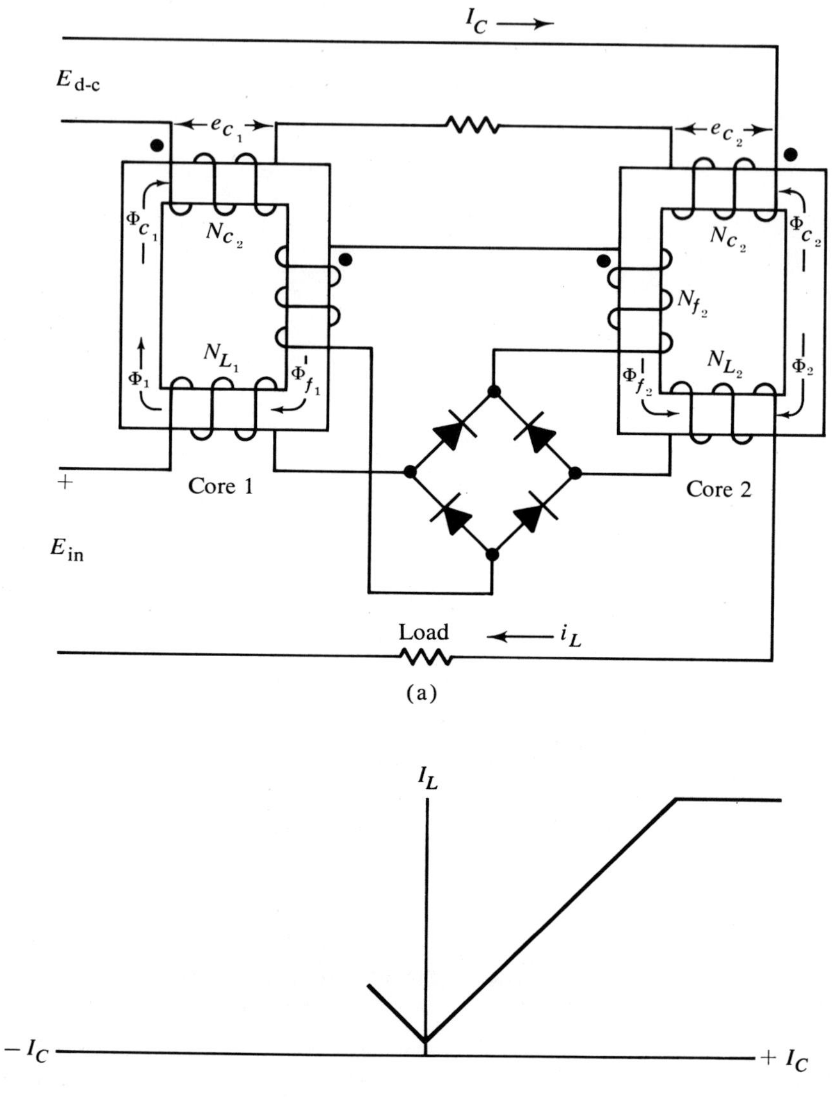

FIG. 3-11. (a) Magnetic amplifier with external feedback; (b) Control characteristics.

that is in opposition to that in the other winding. No net-induced voltage is generated in either winding by the a-c component of the load current, and the feedback current is a function of the load current only.

Figure 3-11 shows the current flux relationships for a part of the cycle when the a-c input voltage has just started to become positive. Core 1 requires less control-current for saturation because the feedback current produces a flux which adds to the control flux.

When core 1 reaches saturation, the relationship of the instantaneous magnetizing forces in core 2 becomes

$$(N_{L2})i_L - (N_{f2})i_f = (N_{c2})i_c \tag{3-35}$$

since $I_L = I_f$

$$(N_{L2} - N_{f2})I_L = (N_{c2})I_c \tag{3-36}$$

This relationship also exists for core 1 during the next half-cycle.

The average control current is related to the average load current, as follows, from Eq. 3-36

$$\frac{I_L}{I_c} = \frac{N_c}{N_L - N_f} \tag{3-37}$$

Equation 3-37 expresses the current gain of the magnetic amplifier after addition of positive feedback.

The power gain may be expressed as

$$W \cong \left(\frac{N_c}{N_L - N_f}\right)^2 \times \frac{R_L}{R_c} \tag{3-38}$$

Equations 3-37 and 3-38 show that both the power gain and the current gains are *increased* by the use of *positive feedback.*

The feedback factor λ is sometimes expressed as the ratio of feedback turns to load turns.

$$\lambda = \frac{N_f}{N_L} \tag{3-39}$$

in which case Eqs. 3-37 and 3-38 become

$$\frac{I_L}{I_c} = \frac{N_c}{N_L(1 - \lambda)}$$

$$W \cong \left(\frac{N_c}{N_L(1 - \lambda)}\right)^2 \times \frac{R_L}{R_c}$$

Negative feedback reduces both power gain and current gain; the equations for each would become, respectively:

$$W \cong \left(\frac{N_c}{N_L(1 + \lambda)}\right)^2 \frac{R_L}{R_c}$$

and

$$\frac{I_L}{I_c} = \frac{N_c}{N_L(1 + \lambda)}$$

Example 3-5 The data for the magnetic amplifier shown in Fig. 3-11 are given as

$$N_{c1} = 1000 \text{ turns}$$
$$N_{f1} = 50 \text{ turns}$$
$$N_{L1} = 200 \text{ turns}$$
$$R_L = 10 \text{ ohms}$$
$$R_c = 25 \text{ ohms}$$

For this amplifier with positive and negative feedback, calculate the:

1. current gain
2. power gain

1. To calculate current gain with positive feedback,

$$\frac{I_L}{I_c} = \frac{N_c}{N_L(1-\lambda)}$$

$$\lambda = \frac{N_f}{N_L} = 0.25$$

$$\frac{I_L}{I_c} = \frac{1000}{200(0.75)} = 6.66$$

and current gain with negative feedback,

$$\frac{I_L}{I_c} = \frac{N_c}{N_L(1+\lambda)}$$

$$\frac{I_L}{I_c} = \frac{1000}{200(1.25)} = 4$$

2. To calculate the power gain with positive feedback

$$W \cong \left(\frac{N_c}{N_L(1-\lambda)}\right)^2 \frac{R_L}{R_c}$$

$$W \cong 17.85$$

and power gain with negative feedback

$$W \cong \left(\frac{N_c}{N_L(1+\lambda)}\right)^2 \frac{R_L}{R_c}$$

$$W \cong 6.4$$

3-11 MAGNETIC-AMPLIFIER APPLICATION

The magnetic amplifier is extremely useful for controlling large amounts of power by means of low-level control signals. One such application is illustrated by the furnace-control circuit shown in Fig. 3-12.

This type of application is ideally suited for the magnetic amplifier because response-time is not a critical factor. Reliable, rugged controls are required for this application; weight and size of the control unit are considered secondary factors.

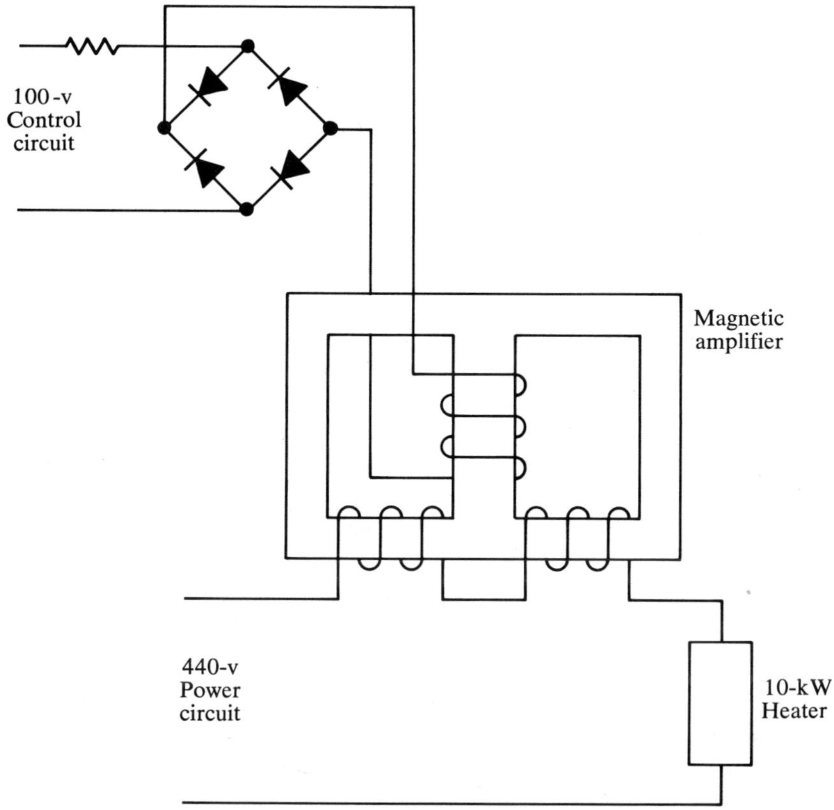

FIG. 3-12. Magnetic amplifier used to control electric furnace.

Calculation of the windings required on the magnetic amplifier for control of the furnace represented by Fig. 3-12 follows. This furnace requires 10 kW of power to achieve maximum rate of temperature rise, and 1 kW will maintain the desired temperature within the furnace. Therefore, the magnetic amplifier must be able to control 9 kW of power. When 10 W of control power are available, the magnetic amplifier must have a power gain of 900. With 100 v applied to the control winding, the control current would be 0.1 A.

The load current, calculated for the 10-kW load at 440 v, is 22.7 A. When the control current is at its maximum, the load current is at its peak value; therefore, the ratio of control-winding turns to load-winding turns may be computed from the relationship

$$\frac{I_L}{I_c} = \frac{N_c}{N_L}$$

$$\frac{22.7}{0.1} = 227$$

Assuming 50 turns for the load winding, the control winding must have 12,500 turns.

To compute the maximum allowable resistance of the control winding, Eq. 3-27 is introduced.

$$W \cong \left(\frac{N_c}{N_L}\right)^2 \times \frac{R_L}{R_c} \qquad R_L = \frac{440 \text{ v}}{22.7 \text{ A}} = 19.5\ \Omega$$

$$900 = (227)^2 \times \frac{19.5}{R_c}$$

$$R_c = 1{,}120\ \Omega$$

Therefore, the control winding must have 12,500 turns with a maximum resistance of 1,120 ohms.

If a portion of the load current is fed back into this magnetic amplifier from a current transformer, as shown in simplified form in Fig. 3-13, the control power required can be reduced.

If the number of turns on the feedback winding is equal to the number of turns on the load winding, and 20 percent of the load current is fed back into the amplifier, then from Eq. 3-35

$$N_LI_L - N_FI_F = N_cI_c \qquad N_L = N_f; \quad I_F = 0.2I_L;$$

Rewriting,

$$N_LI_L - (N_L)0.2I_L = N_cI_c$$

and

$$I_L(N_L - 0.2N_L) = N_cI_c$$

Then

$$\frac{I_L}{I_c} = \frac{N_c}{0.8N_L}$$

If $N_c = 12{,}500$ turns and $N_L = 50$ turns, the new current gain becomes

$$K_c = \frac{12{,}500}{0.8(50)} = 313$$

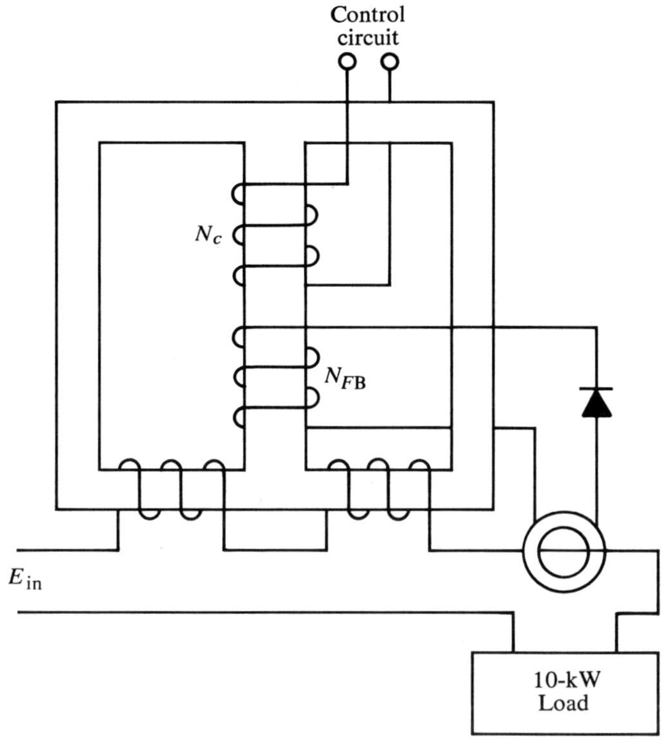

(a)

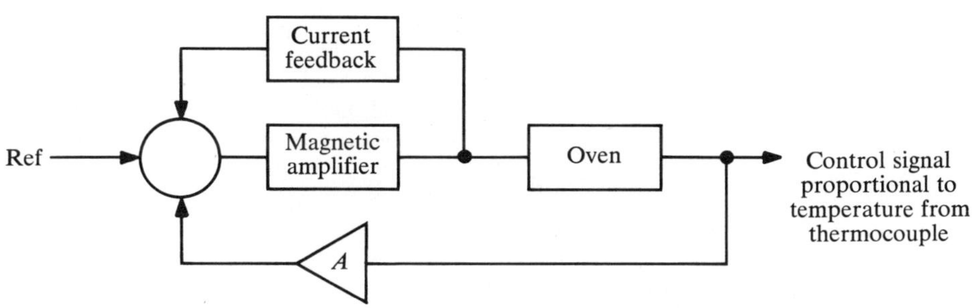

(b)

FIG. 3-13. (a) Magnetic amplifier using current feedback; (b) Block diagram of magnetic-amplifier control circuit.

The power gain is increased and becomes

$$W \cong (313)^2 \frac{19.5}{1120}$$

or

$$W \cong 1710$$

The control power required to regulate this furnace becomes

$$P_c = \frac{9000}{1710}$$

or

$$P_c = 5.27 \text{ W}$$

The 20 percent feedback factor has allowed the reduction of the control power requirement to almost 50 percent of that required without feedback.

Figure 3-13(b) shows a block diagram of a magnetic amplifier used in an automatic temperature regulating circuit. In this application a thermocouple is used to produce a voltage proportional to the temperature of the furnace. After amplification, this voltage is fed back into a negative feedback winding on the amplifier core. The polarity of the feedback current is such that, as the temperature increases, the d-c control-ampere turns are decreased and the current to the heating elements is reduced. The negative feedback from the temperature-sensing device provides automatic temperature control in the manner described.

3-12 CASCADING MAGNETIC-AMPLIFIERS

Section 3-8 stated that the time constant of a magnetic amplifier increases as the power gain is increased. However, improvement in the ratio of time constant to gain can be obtained by the cascading of magnetic amplifiers. When the amplifiers are cascaded, the time constant becomes approximately the sum of the time constants of the individual stages; at the same time power gains multiply. The total time-constant for N stages may be expressed as

$$T_N = NT_1$$

or the time constant for N stages is equal to N times the time constant of a single stage.

The power gains multiply and

$$W_N = W_1^N$$

Therefore, the ratio of time constant to gain improves because the increase in the overall gain is larger than that in the time constant. Cascading cannot be carried on indefinitely because a point of diminishing returns is eventually reached.

Example 3-5 A magnetic amplifier has a time constant of 0.5 sec and a power gain of 500. It is desired to keep the same power gain while reducing the time constant by means of cascading two amplifiers. Assuming the time constant to vary in direct proportion to the power gain, what will be the new time constant of the two magnetic amplifiers in series?

If the power gain of each amplifier is the same, the product of their gains is written as W^2.

Then,

$$500 = W^2$$

and

$$22.5 = W$$

the gain of each series magnetic amplifier.

The time constant of the magnetic amplifier is equal to the product of a constant K and power gain W.

Thus,

$$T_c = KW$$

For the original magnetic amplifier

$$0.5 = K\,500$$

and

$$K = 0.001$$

The constant K is of the same value for the new magnetic amplifier. The new time constant is

$$T = 0.001(22.5)$$

$$T = 0.0225 \text{ sec}$$

The time constant for two cascaded magnetic amplifiers is $T = 2 \times (0.0225) = 0.0450$ sec. and the power gain becomes

$$W = (22.5)^2$$

or

$$W = 500$$

An improvement in the time constant of 11-to-1 has been accomplished by means of cascading, and the overall power gain is still maintained.

3-13 OPERATIONAL CHARACTERISTICS

Many features of the magnetic amplifier make it desirable as the controlling component in a power supply. Some of these features are: extreme ruggedness, power-handling capacity, ease of control, and lack of moving parts. When the device, which is completely static, is properly designed, its life is almost unlimited.

The limitations of the magnetic amplifier are: temperature limitations owing to the rectifiers or diodes, long-time constants, and nonsinusoidal output voltage.

The magnetic amplifier is finding wide acceptance in industrial control applications and is widely used as the control element in regulated-power supplies. There is no doubt that the use of the magnetic amplifier will continue to expand as improvements in materials and designs are made.

Problems

1. Calculate the inductance, in henrys, of a 1000-turn, iron-core coil 10 in. long with a cross-sectional area of 2 in.^{2} The permeability of the iron-core coil is 2000, and the flux density is 100×10^3 lines/in.^{2}
2. Assuming a saturation flux density of 70,000 lines/in.^{2}, for an iron core, calculate the peak value of a 60-Hz voltage that will saturate this core. Assume the core to have a load winding of 200 turns and an area of 1 in.^{2}
3. A 5-in.-long core with a 2-in.^{2} area is wound with 100 turns. If the core material has a μ of 5000 and this μ remains constant, how many lines of flux will be established in this core when 0.5 A is passing through the coil, and when 2 A are flowing in the coil? What will the inductance of the core be when the coil is carrying 0.5 A? 2 A?
4. Calculate the firing angle for the core shown in Fig. 3-2 when $B_s = \dfrac{200\text{v}}{\omega N}$ and E_m equals: (a) 50 v, (b) 100 v, and (c) 150 v.
5. Using the method of Example 3-1, calculate the firing angle of a core composed of material that is biased at a flux density of: (a) 60,000 lines/in.^{2}, and (b) 35,000 lines/in.^{2} (Assume B_s = 70,000 lines/in.^{2})
6. Calculate the losses in the load circuit of the reactor core shown in Fig. 3-6. The control winding has a resistance of 25 ohms and consists of 2000 turns. The load winding has a resistance of 1.5 ohms and consists of 400 turns. Load current just prior to saturation is 8 A.

7. What will be the a-c current and losses in the control winding for the reactor of Problem 6 when the load current is 8 A?
8. Calculate the current gain of a magnetic amplifier having 20 turns on the load winding and 3000 control-winding turns.
9. When the control winding of the reactor in Problem 8 has a resistance of 100 ohms, and a 5-ohm load is connected to the reactor load winding, what will be the power gain of the reactor?
10. Calculate the time constant for the reactor of Problem 8 when it is used in a 60-Hz circuit.
11. The magnetic amplifier of Fig. 3-11 has the following characteristics: $N_c = 2{,}000$ turns, $N_F = 200$ turns, $N_L = 400$ turns; load winding resistance 0.5, $R_{\text{load}} = 5\ \Omega$, $R_c = 40\ \Omega$; and 60-Hz input frequency. Calculate for: a) positive and b) negative feedback:

    1. losses in the load winding when 3 A are flowing through the load and the core is not saturated
    2. losses in the control winding when the load current is 3 A
    3. current and power gain of this reactor
    4. time constant of this reactor

12. Calculate the power required to control a 60-Hz, 440-v, 20-kW heating system by means of a magnetic amplifier with 20 load turns and 1,000 control turns. The resistance of the control winding is 150 ohms. Find the time constant of the required magnetic amplifier.
13. Two magnetic amplifiers are to be used in series. If the time constant and power gain of the first amplifier are 0.2 sec and 100, respectively, and the time constant and power gain of the second amplifier are 0.1 sec and 200, respectively, what will be the time constant and power gain of the series combination?

Questions

1. How is the inductance of an iron-core coil controlled in a saturable reactor?
2. How is the magnetic-flux density in an iron core *reset* when the excitation current is removed?
3. Would a magnetic material with a constant permeability be useful in a saturable reactor core?
4. How will the inductance of an iron-core coil vary during one-half cycle, if the sine-wave input voltage can drive the core beyond saturation? Sketch a curve of X_L vs degrees.

5. Why does the inductance of an iron-core coil remain above zero in the saturation portion of the magnetization curve?
6. What are the advantages of a saturable reactor over a rheostat for controlling the voltage across a load.
7. How will an induced a-c current in the control winding of a saturable reactor affect the amount of power required to control the reactor? Why?
8. What two factors determine the firing angle of the input wave in a saturable reactor?
9. Give two reasons for the resistance of the control circuit of a saturable reactor or magnetic amplifier to be kept as low as possible?
10. Explain the difference between a polarized and nonpolarized reactor.
11. Why is the inductance of a core reduced to nearly zero when a diode is placed in series with an inductor and a load?
12. Explain how the control current can *reset* the core of an inductor which is connected through a diode to a load.
13. In the basic magnetic-amplifier shown in Fig. 3-10 why must the two control windings be wound in opposite directions?
14. For the magnetic amplifier of Fig. 3-10, explain why the inductance of core 1 is reduced to its minimum value as soon as core 2 fires.
15. Define the power gain of a magnetic amplifier.
16. How will the power gain of a magnetic amplifier be affected if the turns on the control winding are increased?
17. How will the time constant of a magnetic amplifier be affected if the load turns on the core are increased?
18. How will negative and positive feedback affect the time constant of a magnetic amplifier?
19. What are the advantages of using positive feedback in a magnetic amplifier?
20. How will the time constant of a magnetic amplifier be changed if the power gain of the device is increased?
21. What are the advantages of cascading magnetic amplifiers?

Gaseous-Tube Circuits

4

The gas-filled tube is used extensively in electronic control circuits. Although semiconductors may replace the gas tube in some applications, the gas tube has advantages over semiconductors in other areas. An understanding of the way in which current is conducted through a gas is necessary, because the operation of many new electronic devices depends upon conduction of current through a gas.

Normally, gas is an insulator and will not conduct. However, when gas molecules are energized by an external source of energy, some electrons become separated from the gas molecules, and the result is that the gas is ionized. Free electrons with negative charges and gas molecules with a positive charge exist in the ionized gas. The presence of these electrons and ions causes the gas to become a conductor. Control of the numbers of free electrons and ions in a gas permits regulation of the electrical characteristics of a gas-filled tube.

4-1 CURRENT-CONDUCTION IN A GAS

A small number of ions and free electrons are always present in a gas enclosed at low pressure between the electrodes of a tube, as shown in Fig. 4-1(a). The gas is ionized because of the bombardment by cosmic rays or ultraviolet radiation, and the ions and electrons travel along random paths within the tube. However, when a low potential is applied to the electrodes of the tube, the ions and electrons are attracted to these electrodes, and a small current is established in the circuit. As the potential is increased, the free charges formed by the radiation are attracted to the electrodes in increasing numbers, and the current increases linearly, as shown in the region 0 to *A* in Fig. 4-1(b). Between points *A* and *B*, Fig. 4-1(b), saturation has been reached, and all electrons and ions released by radiation are attracted to the electrodes. If the radiation were increased in this area, the current would increase; if all radiation were removed, the current would decrease to zero. In this region the current is *not self-maintaining*. When the potential is increased further to point C, some free electrons will gain sufficient energy to ionize the gas atoms by

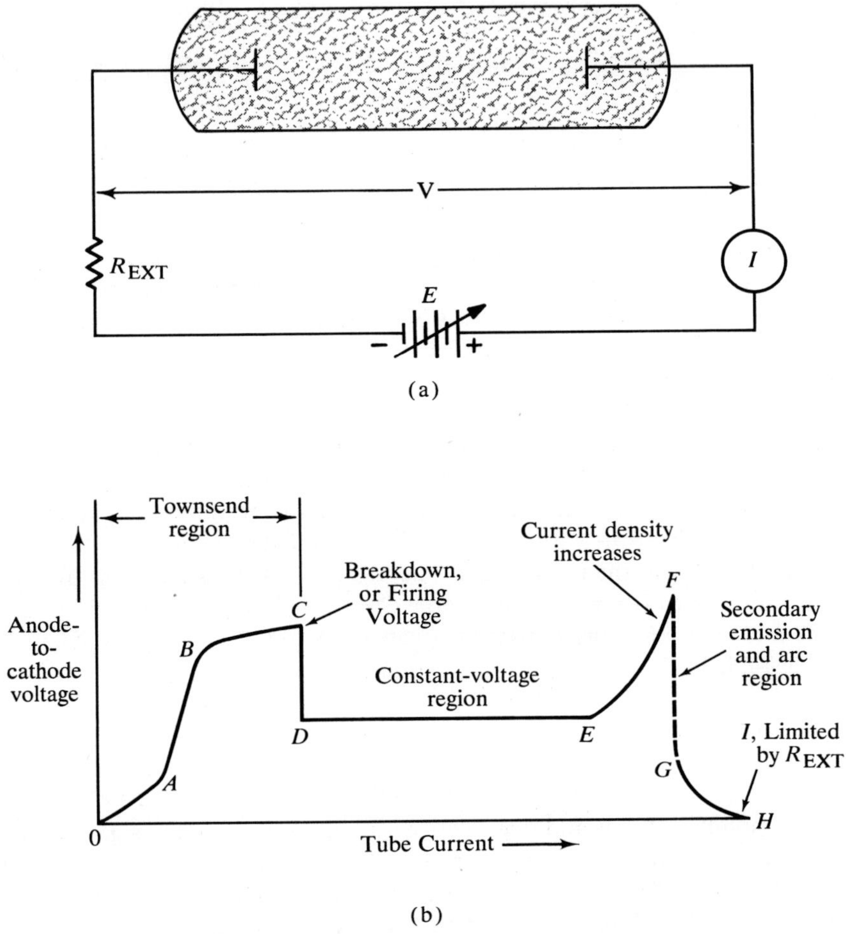

FIG. 4-1. Characteristic curve of voltage-drop variations across gas-filled tube with changes of current.

collision. The further increase in available ions and electrons produced by these collisions results in an increase in tube current up to point C in Fig. 4-1. The portion of the curve from 0 to C is *not self-maintaining*. Current in this portion of the curve depends upon the creation of ions by external radiation.

This portion is called the "Townsend discharge region," after J. S. Townsend, who first investigated this phenomenon.

At a potential just above point *C*, the energy accumulated by the heavy ions traveling through the gas is sufficient to cause secondary emission of electrons when the heavy ions strike the cathode of the tube. These secondary electrons then ionize other gas atoms by collision, and current through the tube is sustained without the aid of external radiation. The current is *self-maintaining* at this point. The voltage at which current becomes self-sustaining is called the *break-down* or *firing potential.*

At this break-down potential the voltage across the tube drops to point *D*, and a visible glow, caused by the secondary emission of electrons, appears at the cathode. The voltage drop across the tube remains constant as the current varies between points *D* to *E*. This constant voltage characteristic makes the gas tube useful as a voltage-regulator tube. As the current increases from point *D* to point *E*, the glow at the cathode covers an increasing portion of the cathode area.

After the entire cathode is glowing, a further increase in current will result in an increase of the current density at the cathode. In the region *E* to *F* the current density at the cathode is increasing, and the voltage drop across the tube also increases.

When the current is increased beyond point *F*, the bombardment of the cathode becomes intense enough to heat the cathode to a point at which thermal emission occurs. At that time the voltage across the tube will begin to fall, and an *arc* will form between the cathode and the anode. From point *G* to point *H* the current within the arc and is limited only by the external circuit resistance. Without external resistance, the current would rise without hindrance, because of the negative resistance characteristics of the arc, until a tube failure or circuit failure occurred.

4-2 THE GAS TUBE AS A VOLTAGE-REGULATOR

The gas tube may be used in a voltage-regulating circuit when it is operated in the constant-voltage section of the volt-ampere curve, shown in Fig. 4-1(b). The voltage across the tube will remain constant while the current varies from point *D* to point *E*, and it is this constant-voltage characteristic that allows use of the glow tube as a voltage-regulator. In Fig. 4-2, the volt-ampere characteristics of a VR or voltage-regulating tube, together with a simplified regulating circuit, are presented.

If the VR tube is to maintain the load voltage at 100 v, the values of the series resistor R_s and of load resistor R_L must be selected so that the glow

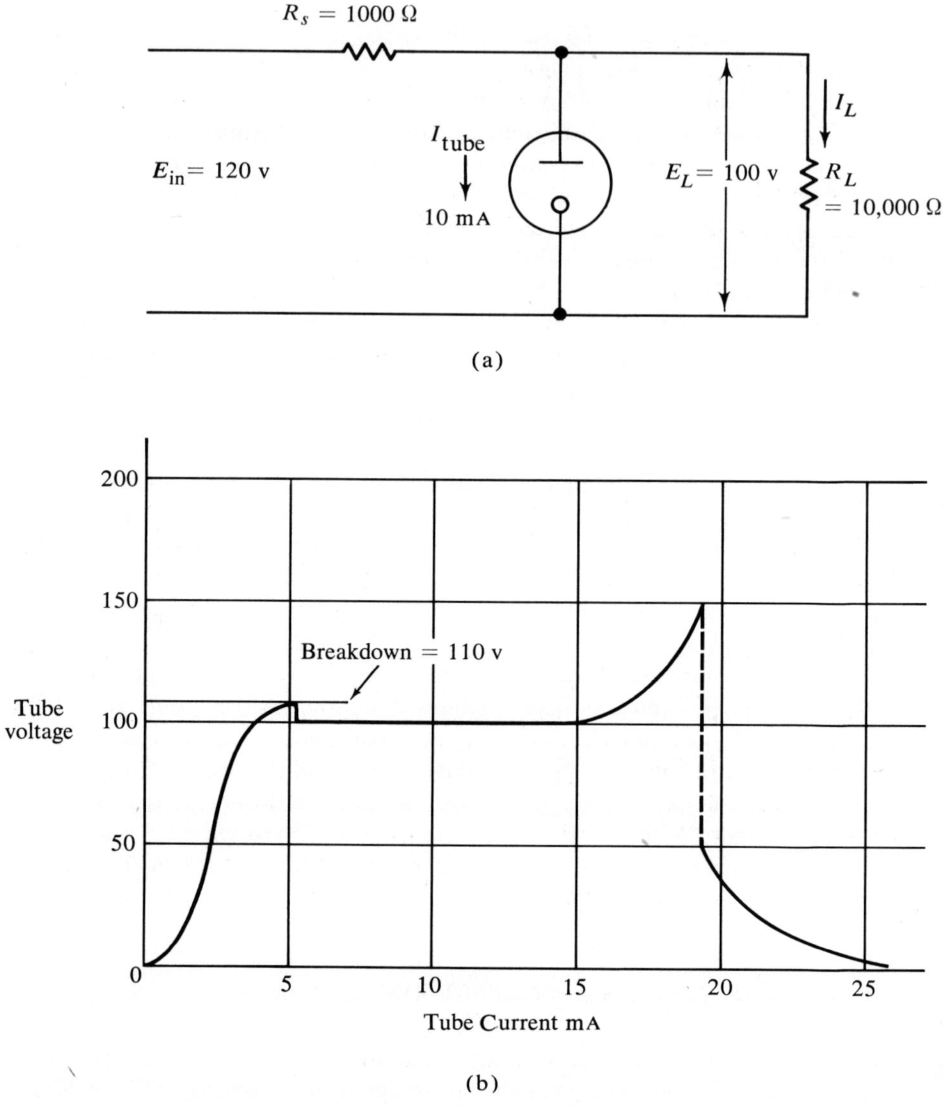

FIG. 4-2. (a) Regulating circuit using gas-filled tube; (b) Typical voltage-current characteristics of VR tube.

tube will operate within the range of 5 to 15 mA. Within that current range the tube voltage-drop remains constant. The E_L is to remain at 100 v and I_L is equal to 10 mA. The value of R_s can be determined if we assume the VR tube will be passing 10 mA when V_L = 100 v and E_{in} = 120 v. Then

$$R_s = \frac{E_{\text{in}} - E_L}{I_L + I_{\text{tube}}}$$

or

$$R_s = \frac{120 - 100}{10\text{ mA} + 10\text{ mA}} = \frac{20\text{ v}}{20\text{ mA}} = 1000\,\Omega$$

With this value of R_s, the VR tube can maintain the load voltage at 100 v despite variation in the load resistor or input voltage.

The VR tube will not fire when the voltage in the circuit of Fig. 4-2(a) is increased from zero and the input voltage is 100 v. The voltage drop in R_s reduces the tube voltage to a value below 100 v. Input must be high enough to compensate for this drop before the tube will fire.

By starting with E_{in} equal to zero and slowly increasing the input voltage until the VR tube fires, it is possible to find the value of E_{in} at which the tube would fire.

Just prior to firing $I_{\text{tube}} = 0$ and

$$E_L = E_{\text{in}} \frac{R_L}{R_L + R_s} \tag{4-1}$$

E_L must equal 110 v for the tube to fire.

$$E_{\text{in}} = E_L \frac{R_L + R_s}{R_L}$$

$$E_{\text{in}} = 110\text{ v}\left(\frac{11\text{ k}}{10\text{ k}}\right)$$

$$E_{\text{in}} = 121\text{ v}$$

minimum value of E_{in} at which the tube will fire.

Example 4-1 Over what range of input voltage will the voltage across R_L be equal to 100 v? Assume the VR tube to have the characteristics shown in Fig. 4-2(b).

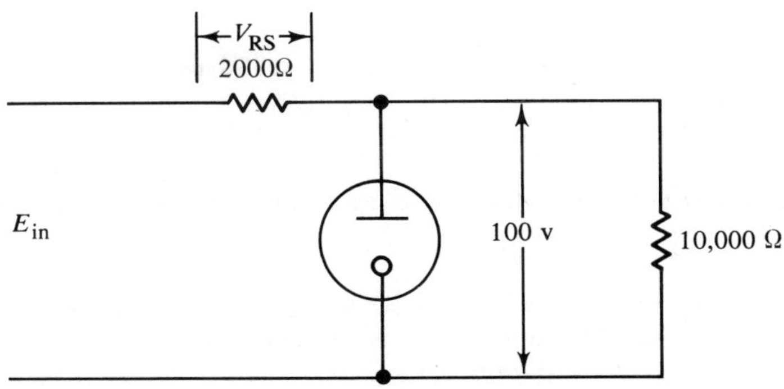

FIG. E4-1. Drawing for Example 1.

1. Find the minimum input voltage required to fire the VR tube. From Eq. 4-1

$$E_{in} = E_L\left(\frac{R_L + R_s}{R_L}\right)$$

$E_L = 110$ v just prior to firing.

Thus,

$$E_{in} = 110\left(\frac{12{,}000}{10{,}000}\right)$$

and

$$E_{in} = 132$$

The value obtained is the minimum input voltage at which the tube will fire.

2. Find the maximum input voltage at which $E_L = 100$ v. At $E_{in\,(max)}$ the VR tube current will be 15 mA. The voltage drop in R_s can be found from

$$V_{RS} = R_s(I_{tube} + I_L)$$
$$V_{RS} = 2000(0.015 + 0.010)$$
$$V_{RS} = 50 \text{ v}$$

Then,

$$E_{in\,(max)} = V_{RS} + V_L$$
$$= 50 + 100$$
$$= 150 \text{ v}$$

3. Find the minimum input voltage at which the tube will maintain 100 v across the load.

At $E_{in\,(min)}$,

$$I_T = 5 \text{ mA}$$

and

$$V_{RS} = 2000(0.010 + 0.005)$$
$$V_{RS} = 30 \text{ v}$$

Then, $E_{in\,(min)} = 100 + 30 = 130$ v. The regulating range is, therefore, 130 to 150 v, and an input voltage of 132 v is required to fire the tube.

The VR tube is also used as an error-amplifying device in voltage-regulation circuits. The error-amplification circuit is illustrated in Fig. 4-3. Since the VR tube will maintain a constant voltage-drop within its operating range, any change in the input voltage will appear across the resistor R_L.

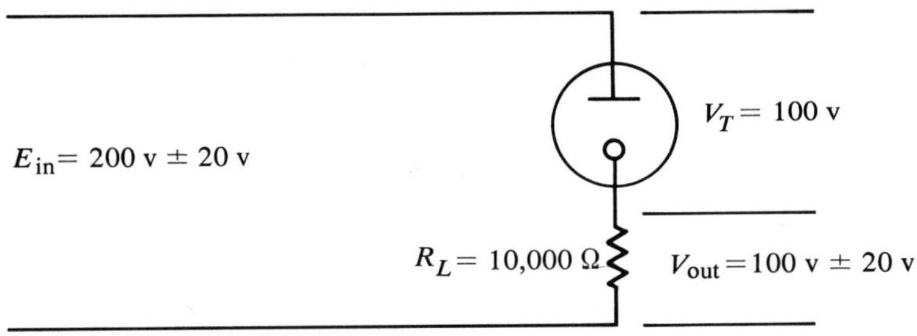

(a)

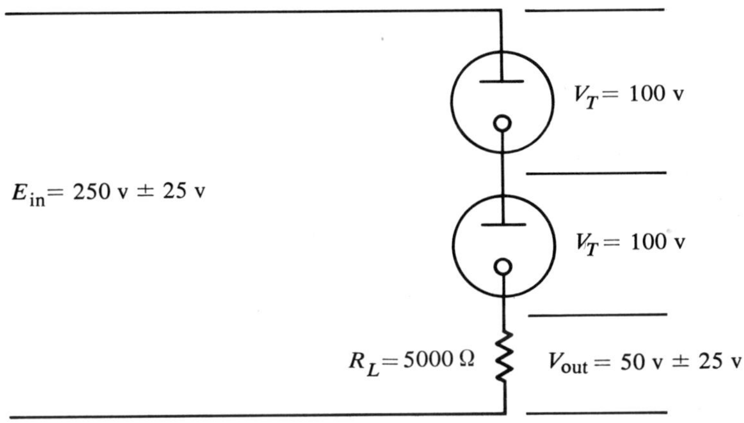

(b)

FIG. 4-3. (a) VR tube used to amplify input-voltage changes; (b) Cascade connection of VR tubes to amplify input-voltage changes.

Example 4-2 The nominal input voltage to the circuit shown in Fig. 4-3(a) is 200 v and the resistor R_L is 10,000 Ω. If the VR tube has the characteristics shown in Fig. 4-2, what will the percentage change in voltage across R_L be if E_{in} changes by 10 percent?

When E_{in} is 200 v, the votlage drop across the VR tube is 100 v, the drop across R_L is 100 v, and I_T = 10 mA. When the input voltage changes by 10 percent or 20 v, the VR tube will maintain its 100-v drop, and V_{out} will change by 20 v (20 percent). The current I_T will also change by 20 percent. This circuit has amplified the change of input voltage by a factor of 2.

Now assume that the input voltage rises to 225 v. The current in the VR tube increases to 12.5 mA, and the drop across R_L increases to 125 v. A higher product for the input change can be obtained by adjusting the value of R_L and by connecting VR tubes in series, as shown in Fig. 4-3(b).

From the above discussion it is clear that the VR tube acts as a load voltage-regulator because of its constant-voltage characteristic. In the design of the regulator circuit, consideration must be given to the fact that the voltage required to start (fire) the tube is greater than the operating (regulating) voltage. For example, the theoretical tube, shown in Fig. 4-2(b), fires at 110 v. Therefore, the values of R_s and R_L must be chosen so that the VR tube will be sure to fire. When the circuit is first energized and I_L is zero, the voltage across the tube must be at least 110 v.

4-3 THE THYRATRON

The thyratron is a hot cathode, gas-filled tube which contains a cathode, an anode, and also a control grid. Figure 4-4 shows the internal structure of the thyratron and a typical grid control characteristic curve. The purpose of the grid is to prevent conduction through the tube until either the anode voltage is raised or the grid potential reduced. If the anode voltage is held constant and the grid voltage reduced below the critical value, an arc forms between the cathode and anode; the current through the tube suddenly increases. The magnitude of the tube current is determined only by the applied voltage and the circuit resistance. After the tube has fired, the grid has no further control over the current.

A grid-control characteristic curve is shown in Fig. 4-4(b). This curve shows each grid voltage required to start an arc for the respective value of anode voltage. This curve reveals that, as the d-c anode voltage is increased, the grid voltage must become more negative in order to prevent conduction. When, for example, 200 v are applied to the anode, the tube will conduct if the grid voltage is less negative than -4 v. Conduction can be prevented by applying a negative voltage greater than -4 v to the grid.

Once the tube fires and the arc is formed between the cathode and anode, the grid cannot stop the current even if the grid is made more negative than the critical value. In order to stop the current, the plate voltage must be reduced to zero until the gas is de-ionized and the grid regains control of the current. The grid will draw current when the tube fires because of the difference in potential between the arc, which contains positive ions, and the grid. Note that care must be taken to limit this current to a safe value.

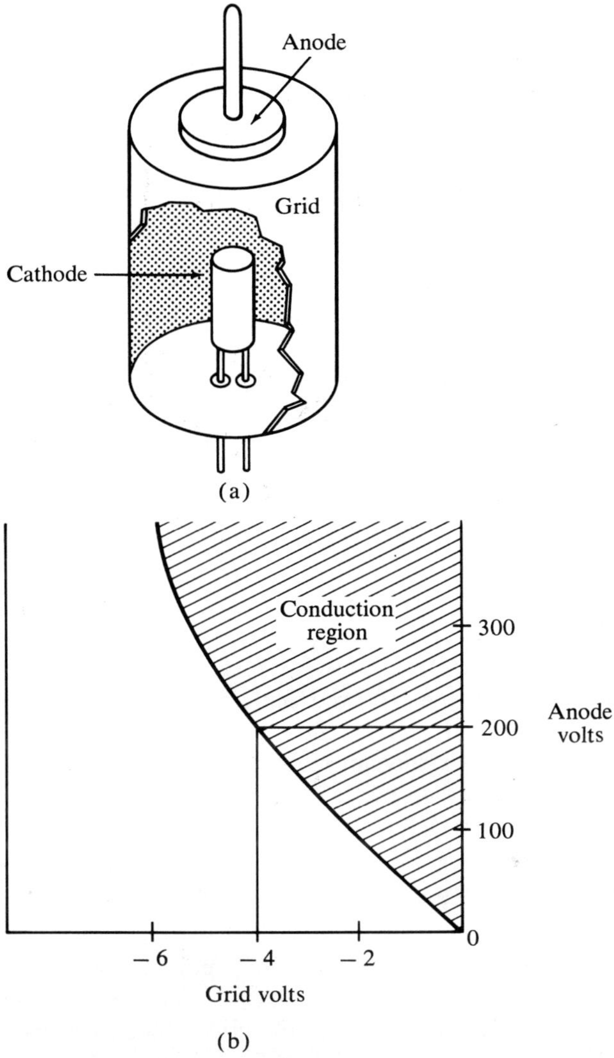

FIG. 4-4. (a) Internal construction of thyratron tube; (b) Anode voltage vs grid-firing voltage for thyratron tube.

4-4 THE D-C THYRATRON CIRCUIT

Thyratron circuits may be classified as either a-c or d-c circuits, depending upon whether the cathode-to-anode voltage is a-c or d-c. In both types of circuit the main problem is the regulation of the anode current. The anode current is controlled by both the grid voltage, which starts the current, and by the anode voltage, which cuts off the current when the latter falls to zero. In a-c circuits, which we shall discuss in a later section, the anode voltage is reduced to zero every half-cycle, and the current is cut off (reduced to zero). In d-c thyratron circuits some auxiliary means of extinguishing the current must be employed. The method of securing arc extinction is the principal problem in d-c thyratron circuits.

A d-c thyratron circuit is used to produce a pulsating d-c voltage whose effective value can be controlled by regulating the width and frequency of the pulses. Such a circuit is referred to as an *inverter* circuit. It provides a method of obtaining large amounts of a-c power from a d-c source.

Figure 4-5(a) illustrates a single-tube inverter circuit which produces a pulsating d-c voltage from a d-c source. When the capacitor voltage is initially at zero and the switch is closed, the voltage across the capacitor will rise along the curve 0 to A shown in Fig. 4-5(c).

Point 3 on the circuit diagram will be at +500 v when the switch is first closed and will become negative with respect to point 2 as capacitor C charges.

The voltage change between points 3 and 2 is shown as curve CD, in Fig. 4-5(c). As the capacitor charges, the anode-to-cathode voltage of the thyratron rises and the negative grid-to-cathode voltage decreases; at point A on the curve, shown in Fig. 4-5(b), the tube *fires* and conduction begins. The thyratron current will be limited by the value of R_1 and R_2.

After the thyratron fires, the capacitor discharges through the thyratron. The anode-to-cathode voltage falls, and the arc is then extinguished. The capacitor then starts to recharge. The value of R_2 must be large enough to prevent the capacitor voltage from rising too quickly. If that voltage rises too fast, the tube refires before de-ionization of the gas occurs, and before the grid regains control.

Example 4-3 A thyratron inverter circuit is required to produce a 60-Hz, a-c voltage with a peak-to-peak value of 300 v from a 500-v, d-c source. Find the values of R_1, R_2, R_3, R_4, and C for the circuit, with tube characteristics as shown in Fig. 4-5.

In the curve of 4-5(b), the grid voltage required to fire the thyratron with an anode voltage of 300 v is seen to be −3 v. When the anode voltage is 300 v, the drop across R_2 is 200 v, and the drop across R_3 must be 197 v for the tube to fire.

If

$$I_1 = 50 \text{ mA (a typical bias current)}$$

$$R_3 = 3940\ \Omega$$

$$R_4 = 6060\ \Omega$$

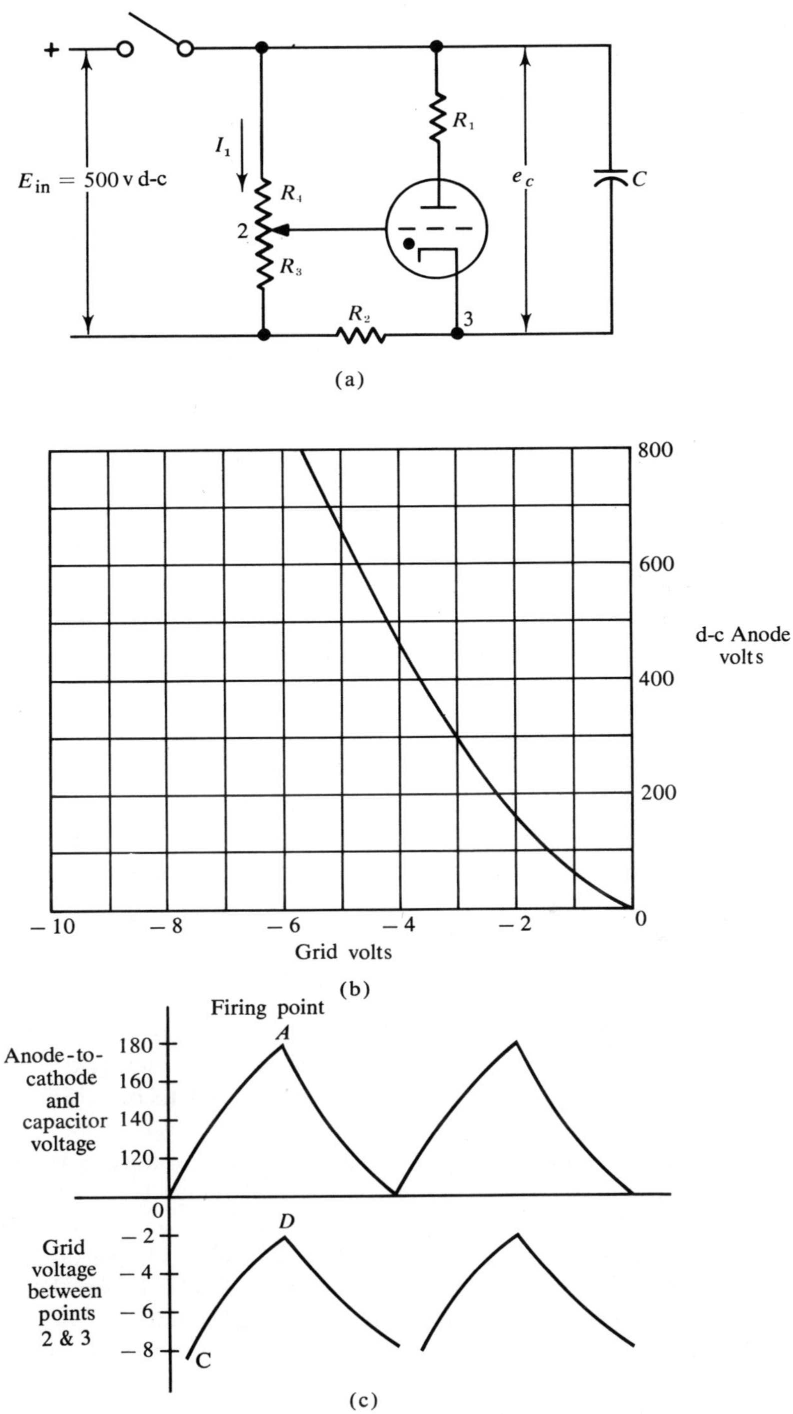

FIG. 4-5. (a) D-C-to-A-C-inverter circuit using thyratron tube; (b) Typical characteristic curve for thyratron tube; (c) Waveforms in thyratron-inverter circuit.

For 60-Hz operation the anode voltage must reach 300 v in 1/120 or 0.0083 sec.

The voltage rise on a capacitor may be found from

$$e_c = E[1 - e^{-t/RC}]$$

$$300 = 500[1 - e^{-x}]$$

$$e^{-x} = 0.4$$

$$x = 0.92$$

$$t/RC = 0.92$$

$$RC = 0.0091$$

Assume, $C = 100\ \mu\text{F}$, then $R_2 = 91\ \Omega$

The charging and discharging periods must be equal if the period for one cycle is 1/60 sec. This requirement yields $R_2 = R_1 = 91\ \Omega$.

A widely used thyratron circuit is illustrated in Fig. 4-6(a). This circuit will produce the waveform shown in Fig. 4-6(c) across the capacitor C. Before attempting to analyze the circuit of Fig. 4-6(a), look at the simplified version of this circuit in Fig. 4-6(b). When S_1 and S_2 are open, there is a negative bias on the tube, which will not conduct. When S_1 is closed, the tube will fire and the current through R_1 will produce a voltage drop across R_1. The capacitor C will charge to this voltage through R_2 with the polarity as indicated. If S_1 is now opened, a negative bias will be applied to the tube but will have no effect upon the conduction of current through the tube.

If S_2 is now closed, the cathode of the tube will be connected to the positive side of the capacitor C, and the anode will be connected to the negative side. This reversal of polarity across the tube will extinguish the arc, and the tube will be cut off. Resistor R_g prevents shorting of the bias battery, and resistor R_2 prevents shorting of the d-c power supply.

The circuit of Fig. 4-6(b) operates in the same manner as that of Fig. 4-6(a) except: (1) switch S_2 is replaced by tube 2, (2) switch S_1 is replaced by a transformer, and (3) the a-c power supply is connected to each grid. Assume an a-c voltage to be applied to the transformer, point 1 to be positive with respect to point 2, and point 3 to be negative with respect to point 2. Tube 1 will conduct, the current through resistor R_1 will produce a voltage drop across R_1, and capacitor C will charge to this voltage level. The capacitor will charge through R_2 with point B positive and point A negative. When the a-c supply reverses, tube 2 will conduct. The positive charge on point B of the capacitor will be connected to the cathode of tube 1, and the tube will be cut off. The grid of tube 1 will, therefore, regain control.

With tube 2 conducting, the capacitor will charge to the opposite polarity

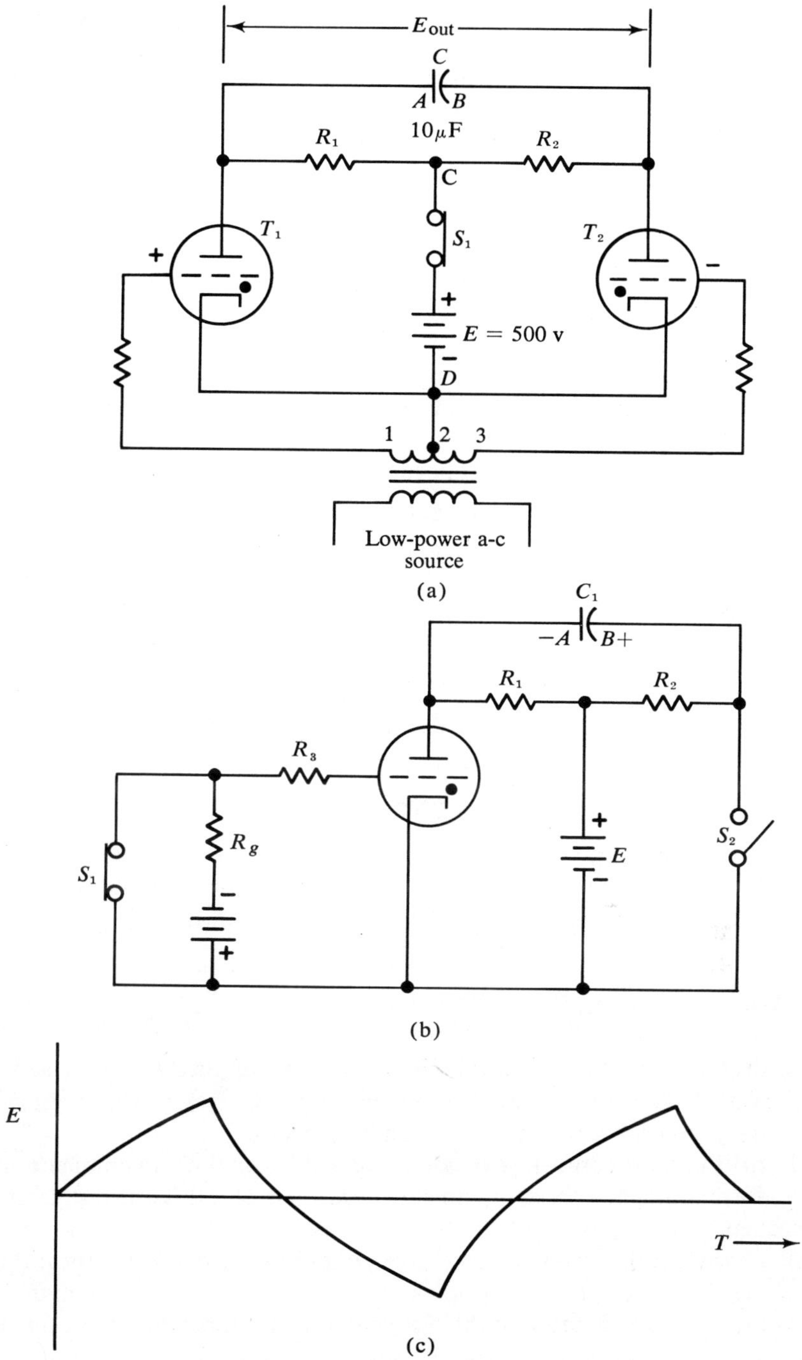

FIG. 4-6. (a) Dual-tube thyratron-inverter circuit; (b) Simplified circuit of dual thyratron-inverter; (c) Output waveform from thyratron-inverter.

with point B negative. After the grid of tube 1 is again driven positive by the a-c input voltage, it will begin to conduct. The charge on the capacitor will extinguish tube 2, and the cycle will repeat. The two tubes will alternate between conducting and nonconducting. When R_1 and R_2 are replaced with the primary of a center-tapped transformer, the circuit can supply large amounts of a-c power from a d-c source.

Example 4-4 The circuit of Fig. 4-6(a) is required to produce a 300 v peak-to-peak a-c voltage across the 10-μF condenser. When the input voltage is 60 cycles, calculate the value of R_1 and R_2.

To obtain a 300-v peak-to-peak voltage across C, the capacitor must charge, within 1/120 or 0.0083 sec, to +150 v on plate A and then to −150 v on plate B.

The rate of voltage rise on a capacitor may be expressed as

$$e_c = E_{\text{in}}(1 - e^{-T/RC})$$

$$150 = 500(1 - e^{-x})$$

$$e^{-x} = 0.7$$

$$x = 0.36$$

$$T/RC = 0.36$$

$$RC = 0.023$$

$$C = 10\ \mu\text{F}$$

and

$$R = 2300\ \Omega$$

The capacitor charges around the loop A, B, C, D, that is, when tube 1 conducts the capacitor charges through R_2, and when tube 2 conducts, through R_1. For equal conduction periods $R_1 = R_2 = 2300$ ohms.

4-5 THE A-C THYRATRON CIRCUIT

In a-c thyratron circuits, in which anode power is supplied by an a-c source, extinction of the anode current is not usually a problem. The method of controlling the grid is the principal area of concern.

The grid may be controlled by an a-c or a d-c signal or a combination of a-c and d-c voltages. These methods of grid control will be investigated in subsequent paragraphs.

All a-c thyratron circuits may be divided into two broad categories: those supplying d-c current to a load or rectifying circuit and those supplying a-c current to the load. Because of the distorted waveform produced, a-c and d-c thyratron circuits are used only where sinusoidal waveforms are not required

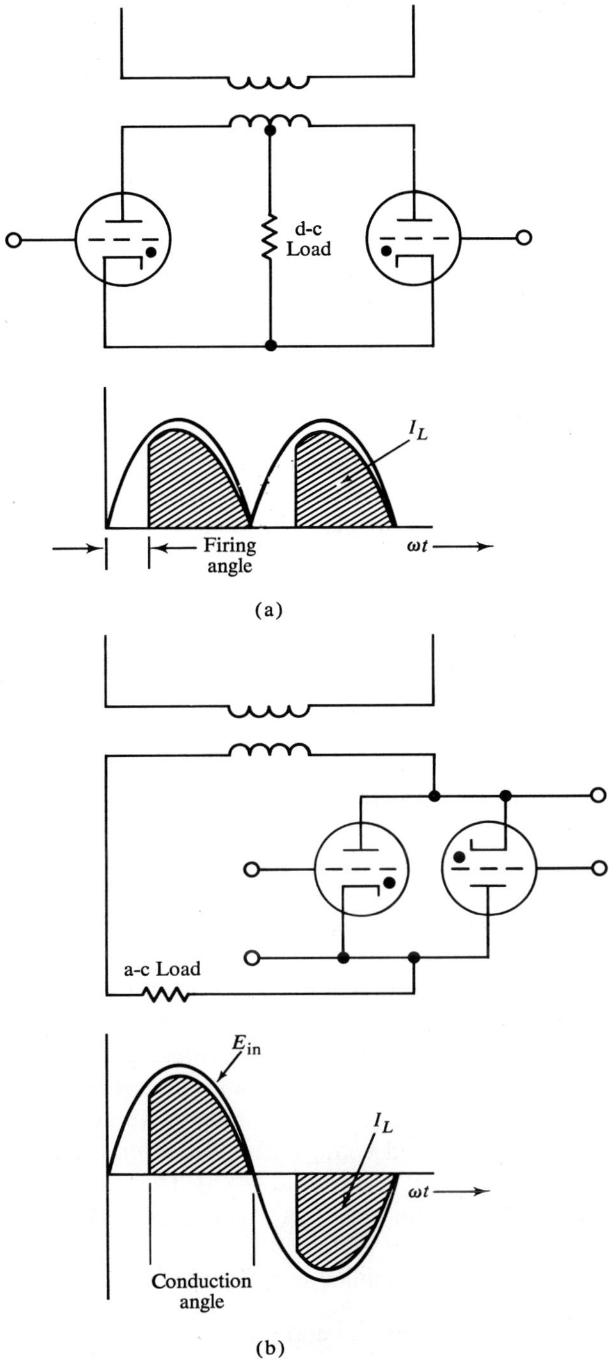

FIG. 4-7. (a) Full-wave thyratron-rectifier circuit; (b) Back-to-back thyratrons for delivering a-c power to a load.

(heating and motor-control applications). Elementary circuit diagrams of a-c thyratrons supplying both a-c and d-c loads appear in Fig. 4-7.

The thyratron circuit of Fig. 4-7(a) can be considered to be a full-wave rectifier in which the diodes have been replaced by the thyratron tubes. Before either thyratron can conduct, its plate voltage must be positive; while the plate is positive, the grid voltage must be of a magnitude that will allow the tube to fire. By controlling the grid voltage, the conduction period of the thyratron can be regulated. When a high negative-voltage is applied to the grids, the d-c load current will be zero. With low negative-voltages on the grids, the thyratron will fire at some low positive-input voltage. The magnitude of the grid voltage will control the angle ϕ, through which the input voltage vector must travel before the thyratrons will fire. The average current in the load can therefore be controlled by regulation of the voltage on the thyratron grids.

The back-to-back circuit of Fig. 4-7(b) operates on the same principle. The conduction angle and, therefore, the average current through each thyratron is controlled by means of the voltage on the thyratron grid.

4-6 THYRATRON CIRCUITS WITH D-C CONTROL

A simple a-c thyratron circuit with d-c control is shown in Fig. 4-8. In this circuit the a-c anode voltage varies as a function of ωt, therefore, the critical grid voltage (firing voltage) constantly varies according to each instantaneous value of anode voltage.

The critical grid voltage is determined by both the control characteristic of the tube and the anode voltage on the tube. A plot of the critical grid voltage vs ωt of an a-c thyratron circuit is a *grid locus* plot. A typical plot is shown in Fig. 4-9(b). This grid locus plot is obtained from the characteristic curve of the tube shown in Fig. 4-9(b). The critical grid voltage, which is negative, is plotted below the horizontal axis and the instantaneous, positive-anode voltage above the axis. For each instantaneous value of positive-anode voltage, the negative-critical grid voltage, obtained from the curve of 4-9(a), is plotted below the anode voltage on curve 4-9(b). The result is the grid locus plot shown. If it were desired to have this particular tube fire at a voltage of 200 v, a grid voltage of -4.25 v, obtained from the curve, is required. The firing angle, when -4.25 v are applied to the grid and $E_m = 400$ v, is found from the following

$$200 \text{ v} = 400 \sin \omega t$$

$$\sin \omega t = 0.5$$

$$\omega t = 30°$$

From the circuit shown in Fig. 4-8 the firing angle α is determined by the d-c grid voltage. When the thyratron fires, the voltage across R_L suddenly rises to the value a, indicated in Fig. 4-8(b) (neglecting tube drop). The load voltage then follows the input waveform from point *a* to point *b*, where the thyratron

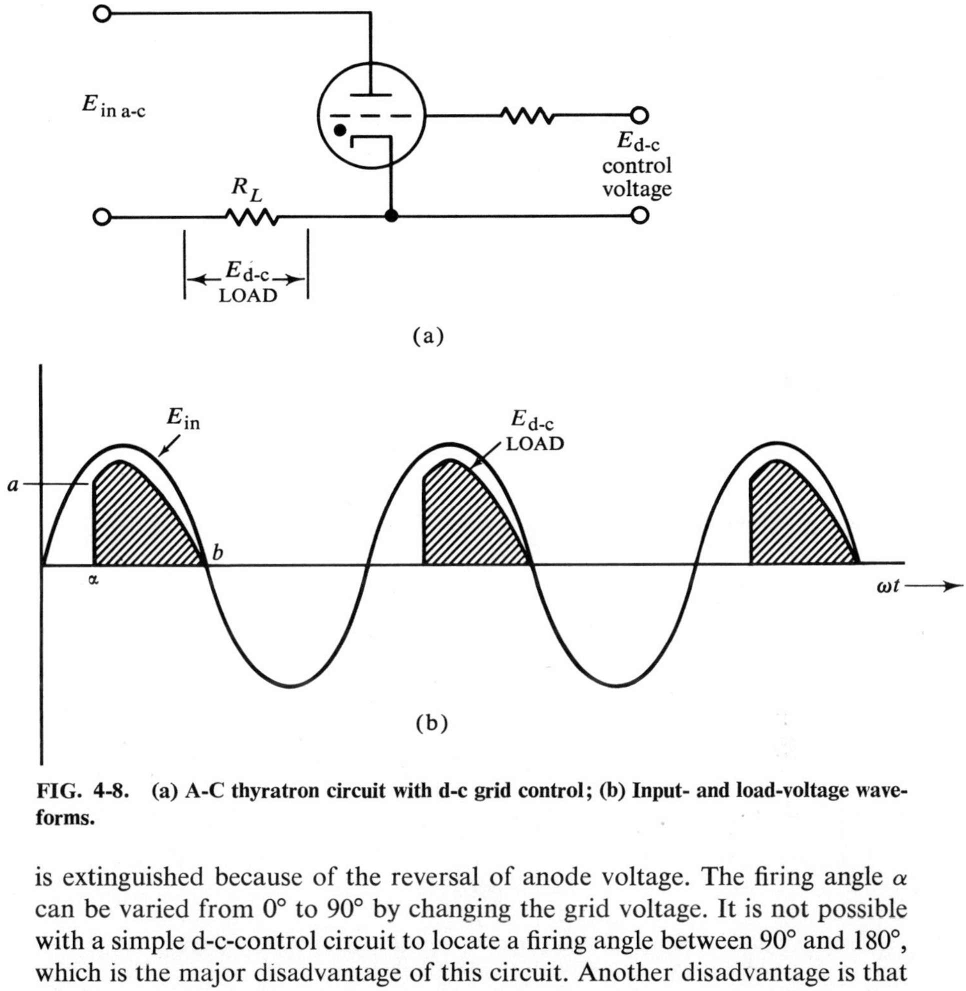

FIG. 4-8. (a) A-C thyratron circuit with d-c grid control; (b) Input- and load-voltage waveforms.

is extinguished because of the reversal of anode voltage. The firing angle α can be varied from 0° to 90° by changing the grid voltage. It is not possible with a simple d-c-control circuit to locate a firing angle between 90° and 180°, which is the major disadvantage of this circuit. Another disadvantage is that of shift in firing angles near 90°, where the angle is likely to shift from cycle to cycle because a slight change in bias voltage can cause a large shift in the firing angle.

Example 4-5 The thyratron represented by the curve in Fig. 4-9 is to be used in a circuit whose maximum voltage is 300 v with -3 v applied to the grid

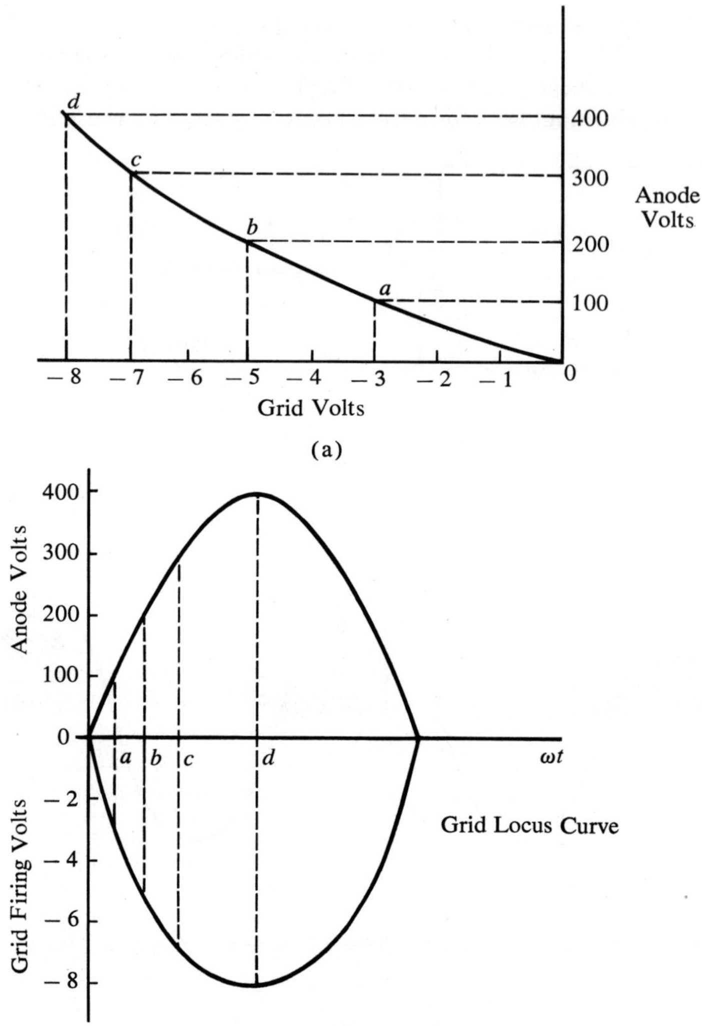

(b)

FIG. 4-9. (a) Thyratron characteristic curve; (b) Grid locus plot of thyratron taken from characteristic curve.

of the tube. What will be the magnitude of the input voltage when the thyratron fires? What will the firing angle and conduction angle be at this point?

Refer to Fig. 4-9(a). With −3 v applied to the grid, the thyratron will fire with 100 v applied to the anode. The firing angle is computed as follows:

$$100 = 300 \sin \omega t$$

$$\omega t = \sin^{-1} 0.333$$

The firing angle ϕ_f is, therefore, 19.5°.

The conduction angle can be computed as

$$\phi_c = 180 - \phi_f$$

$$\phi_c = 160.5$$

4-7 THYRATRON CIRCUITS WITH A-C AND D-C CONTROL WITH A-C AND D-C BIAS

In Fig. 4-10 a circuit is shown which eliminates the disadvantage of the simple d-c-biased thyratron circuit and permits thyratron firing from 0° to 180°.

An alternating voltage has been added in series with the d-c bias; the phase of this voltage is shifted to lag the anode supply voltage by some angle $(90 + \phi)°$. This phase-shift, between the input and grid voltages, is obtained by placing resistor R_1 in series with the primary of the grid voltage transformer. The angle ϕ, between the supply voltage and transformer current, can be controlled by varying the value of R_1. The transformer secondary is connected to the grid and a phase angle having a magnitude of $(90 + \phi)°$ exists between the anode supply voltage and grid voltage. A phasor diagram of the circuit voltages is shown in Fig. 4-10(b).

If -3 v dc is applied to the grid by the battery, and the peak value of the a-c voltage E_s is 3 v; the total grid voltage then will vary from zero to -6 v. The -3 v d-c level is taken as zero reference for a-c voltage in Fig. Fig. 4-10(c), in which the instantaneous grid voltage is seen to vary from zero to -6 v. Whenever the anode voltage is plus and the sum of the d-c and a-c voltages on the grid crosses the grid locus curve, the thyratron will fire.

When the magnitude of the d-c bias is changed, the intersection of the total grid voltage with the grid locus curve will shift, causing a change in the firing angle. This angle can now be shifted from zero, with a positive grid voltage, to approximately 180° with a negative grid voltage. The disadvantage of this circuit is that large changes in the d-c-control voltage are necessary to change the firing angle. However, the angle of intersection between the grid locus curve and the grid control voltage is large, and the firing angle has little tendency to shift.

Example 4-6 If a d-c voltage of -1 v and an a-c voltage of 2 v zero-to-peak were applied to the grid of the thyratron represented in Fig. 4-10(c), at what angle and input voltage would the tube fire? Assume $\phi = 0$. The tube would

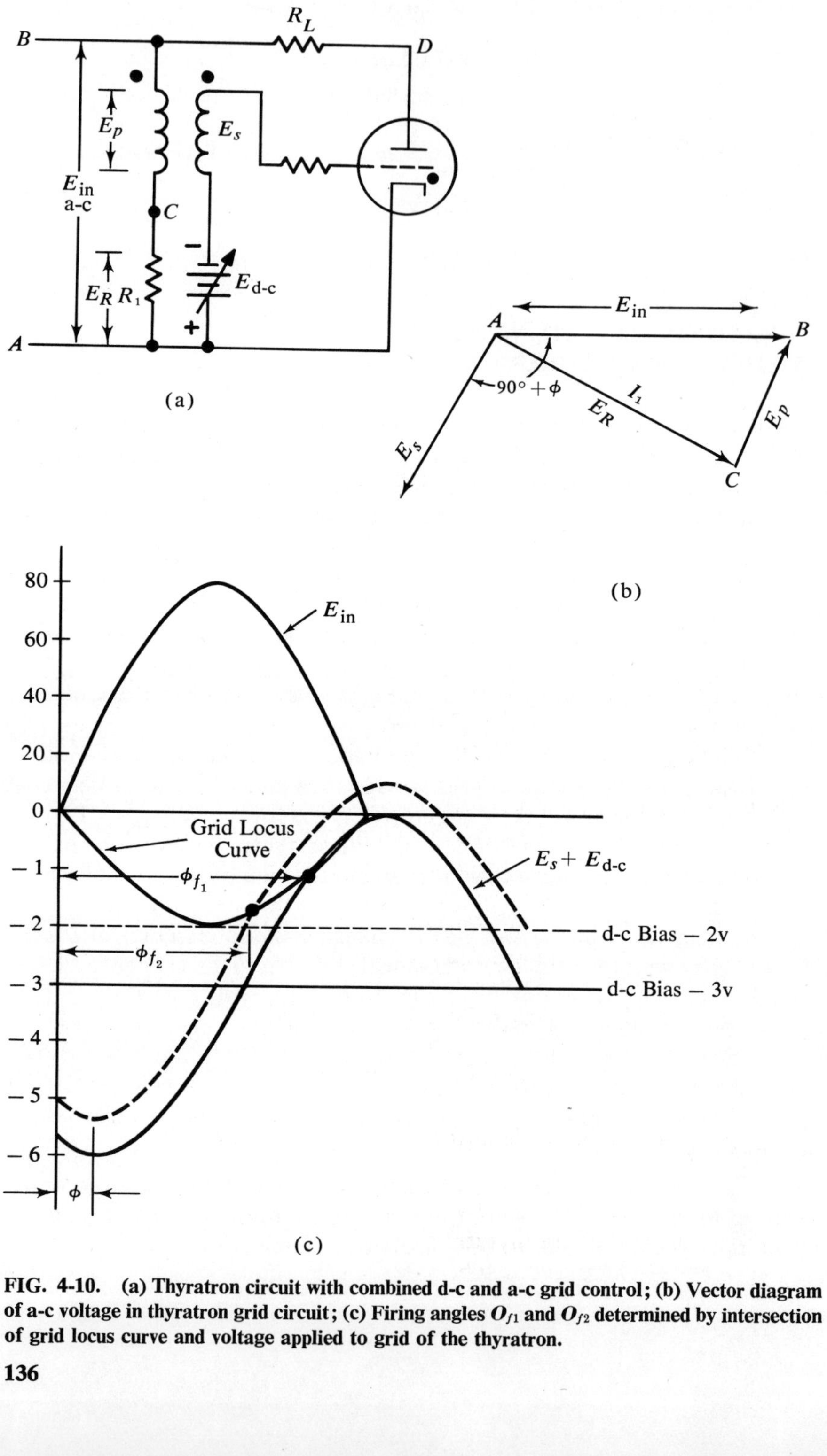

FIG. 4-10. (a) Thyratron circuit with combined d-c and a-c grid control; (b) Vector diagram of a-c voltage in thyratron grid circuit; (c) Firing angles O_{f1} and O_{f2} determined by intersection of grid locus curve and voltage applied to grid of the thyratron.

fire when the grid locus curve intersects the input voltage. The grid locus plot can be described by the equation $e \cong -2 \sin \omega t$. The curves will intersect when

$$\begin{aligned} -2 \sin \omega t &= -1 + 2 \sin (\omega t - 90) \\ 1 &= 2 \sin \omega t + 2 \sin (\omega t - 90) \\ 0.5 &= \sin \omega t + \sin (\omega t - 90) \end{aligned}$$

which may be solved by the identity

$$\begin{aligned} \sin \mathrm{A} + \sin \mathrm{B} &= 2[\sin \tfrac{1}{2}(\mathrm{A} + \mathrm{B})][\cos \tfrac{1}{2}(\mathrm{A} - \mathrm{B})] \\ 0.5 &= 2[\sin \tfrac{1}{2}(2\omega t - 90)][\cos \tfrac{1}{2}(90)] \\ 0.5 &= 2[\cos \tfrac{1}{2}(2\omega t)](\cos 45°) \\ 0.25 &= \cos \omega t(0.707) \\ 0.353 &= \cos \omega t \\ \omega t &\cong 69.3° = \text{firing angle} \end{aligned}$$

The magnitude of input voltage when the tube fires is

$$\begin{aligned} e_{\text{in}} &= E_{\text{m}} \sin \omega t \\ &= 80 \sin 69.3° \\ &= 75 \text{ v} \end{aligned}$$

4-8 THE PHASE SHIFT NETWORK AND THYRATRON GRID CONTROL

It is apparent from Fig. 4-10 that the firing angle of the thyratron can be controlled by shifting the phase angle of the a-c component of the grid voltage. The firing angle can be controlled in this manner even when the d-c component of grid bias is not present. A phase-shift network can provide the required grid control without the d-c bias. A suitable network with a phasor diagram of voltage magnitudes and angles is shown in Fig. 4-11. This network makes use of a center-tapped transformer, adjustable resistor, and fixed reactor. The grid voltage is obtained between points B and D, and the angle between V_{BD} and E_{AC} is adjusted by the slider resistor R. The current in the RL circuit lags the applied voltage by the angle ϕ, as shown in Fig. 4-11(b). The voltage across the resistor V_R lags the applied voltage E_{AC} by the same angle ϕ. The voltage V_L across the reactance X_L leads the current by 90°, and the vector sum of these voltages, V_L and V_R, is equal to the applied voltage E_{AC}. The voltage V_{BD} is then at an angle β with respect to the applied voltage. As the resistance R is varied, the current I also varies, and the angles ϕ and β will change. The input voltage E_{AC} is constant, and V_L will always be at a

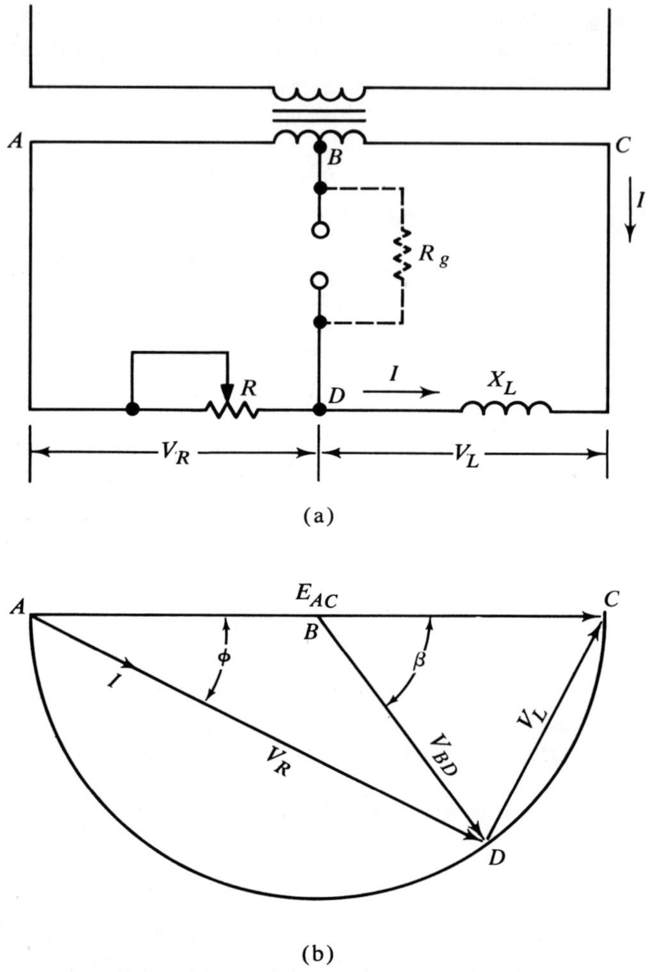

FIG. 4-11. (a) Phase-shifting network for thyratron grid control; (b) Vector diagram of voltages in phase-shifting network.

right angle to V_R; therefore, the point D is located on a semicircle whose center is at point B and whose radius is V_{BD}. The voltage V_{BD} is, therefore, of constant magnitude, and its angle β can vary from 0 to 180° with respect to E_{AC} as R goes from zero to infinity. The angle ϕ is expressed as

$$\phi = \tan^{-1}\left(\frac{\omega_L}{R}\right)$$

$$\beta = 2\tan^{-1}\left(\frac{X_L}{R}\right) \tag{4-2}$$

In this circuit the inductive reactance X_L may be replaced by a capacitive reactance X_c with similar results. In addition, the load resistor R_g must be large compared to R and X_L, or an error will be introduced into Eq. 4-2.

Example 4-7 It is desired to fire the thyratron tube represented by Fig. 4-10(c) at 120°. A center-tapped transformer with a peak-output voltage of 3 v

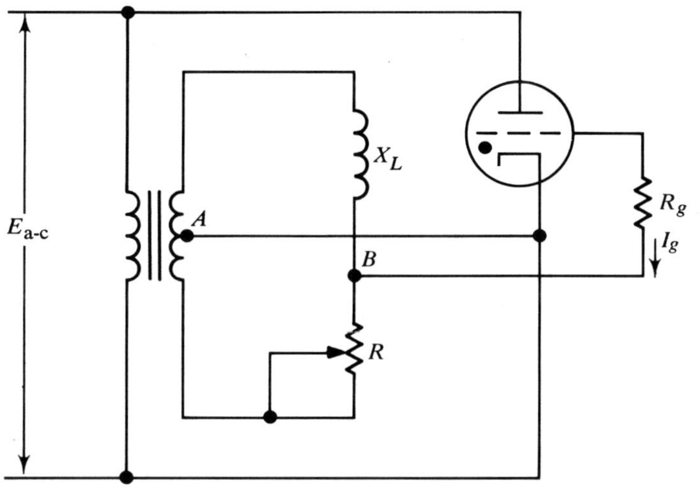

FIG. E4-7. Thyratron tube with phase-shift firing network.

is available. Calculate the values of R and C required for the grid control circuit shown in Fig. E4-7, when the anode of the thyratron is connected to a 60-Hz supply with a peak voltage of 400 v.

From Eq. 4-2

$$\beta = 120° = 2 \tan^{-1} \frac{X_L}{R}$$

$$\phi = 60° = \tan^{-1} \frac{X_L}{R}$$

Resistor R_g is present only to prevent excessive grid current in the control circuit.

To design the phase-shifting network we must assume a value for one component, so let $R = 1$ kΩ.

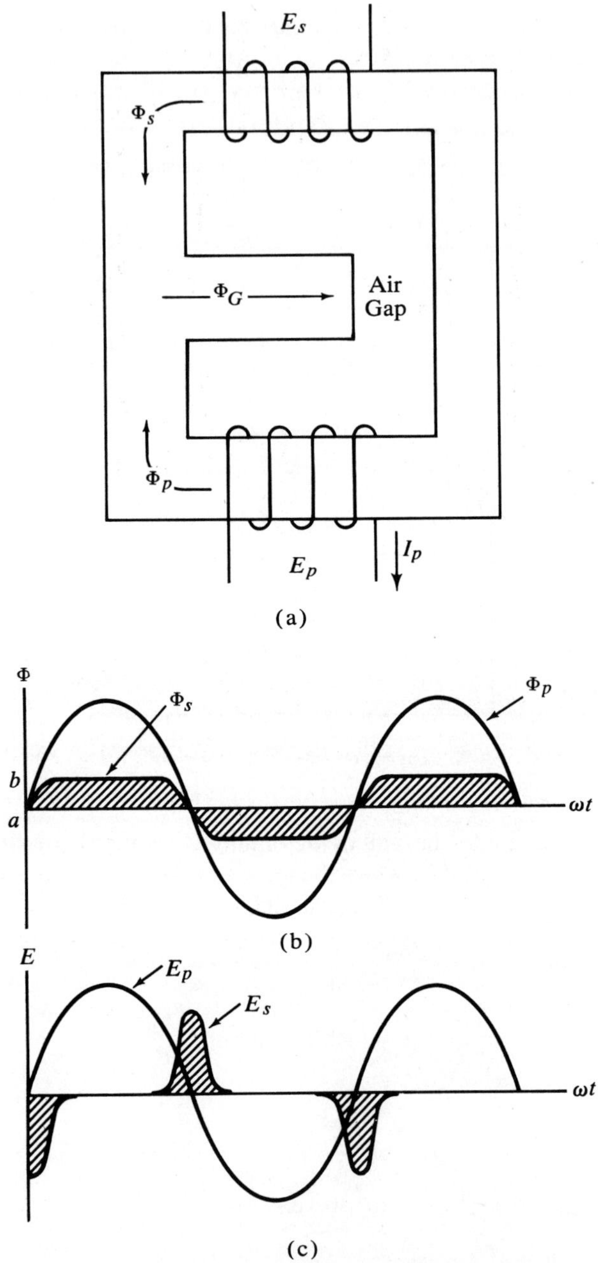

FIG. 4-12. (a) Core- and winding-arrangement of peaking transformer; (b) Flux relationships in peaking-transformer core; (c) Voltage relationships in peaking transformer.

$$\tan 60° = 1.73$$

$$1.73 = \frac{X_L}{R}$$

$$X_L = 1.73 \text{ k}$$

$$L = \frac{1.73 \text{ k}}{2\pi f}$$

$$L = 4.6 \text{ H}$$

4-9 THE PEAKING TRANSFORMER

The peaking transformer is widely used in thyratron grid circuits. This transformer has an output waveform that is sharply peaked and whose duration is usually less than one-half cycle. Figure 4-12 shows a typical core and winding arrangement for a peaking transformer. When an a-c voltage is applied to the primary winding, primary flux density Φ_p is produced, having the sinusoidal waveform shown in Fig. 4-12(b). The primary core does not saturate because of its large cross-section. However, the core of the secondary will become saturated at some flux level Φ_s. During the early portion of the cycle, when the primary flux density is changing from *a* to *b* as in Fig. 4-12(b), a voltage is induced in the secondary winding. At point *b* the secondary core saturates, but no further change of flux linkages in the secondary winding occurs. The voltage induced in the secondary, therefore, falls to zero. After the secondary core is saturated, any further flux produced by the primary ampere-turns spans the air gap and shunts the secondary winding. The output waveform of the secondary voltage is shown in Fig. 4-12(c). The circuit may be used to control the firing voltage on the grid of a thyratron tube if a phase-shifting network is applied to the primary of the peaking transformer.

4-10 NONSINUSOIDAL CONTROL OF THYRATRON CIRCUITS

When an a-c voltage is used to fire a thyratron tube, circuit components or variations in line voltage may cause shifting of the firing angle. This angle can be accurately controlled if the waveform of the voltage applied to the grid is a vertical pulse with a very sharp leading edge. Figure 4-13 shows a circuit with such a pulse, which is nearly vertical when it crosses the control locus. A peaking transformer is used to produce a pulse with a sharp leading

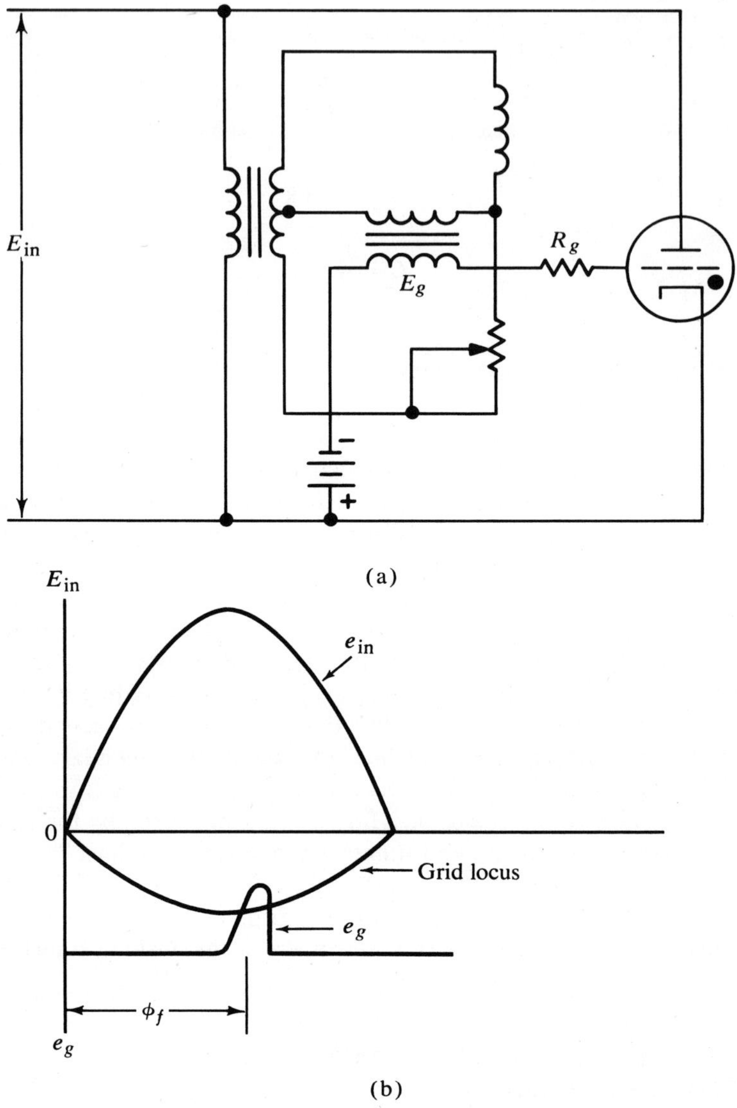

FIG. 4-13. (a) Thyratron tube fired by phase-shifting network with peaking transformer; (b) Grid locus plot and peaking-transformer output intersect at firing angle.

edge. The phase angle of this pulse is controlled by the phase-shifting network in the primary of the transformer.

Another method of obtaining a sharp pulse for the grid of the thyratron is to differentiate the output of a square-wave generator by means of an *RC*

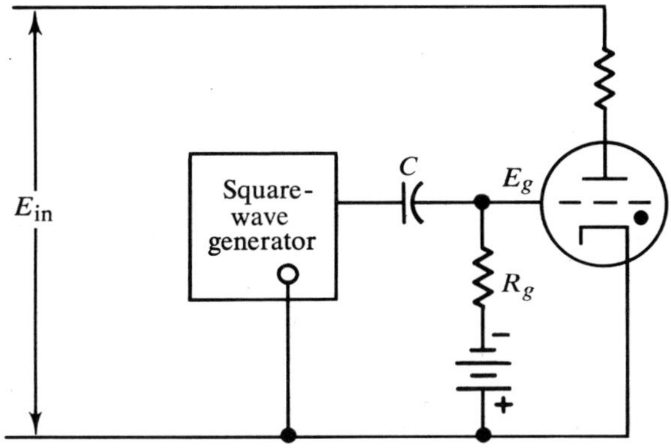

(a)

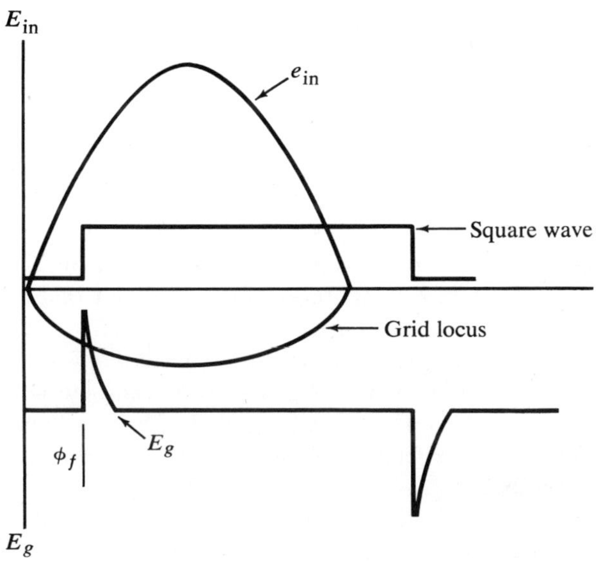

(b)

FIG. 4-14. (a) Thyratron with square-wave generator in firing circuit; (b) Intersection of leading edge on square wave with grid locus plot determines firing angle O_f.

network. This differentiation produces a waveform with a very steep front; this waveform can be used to control the grid of the tube. A simplified version of the circuit is shown in Fig. 4-14.

Problems

1. A VR tube connected as shown in Fig. 4-2(a) with the characteristics illustrated in Fig. 4-2(b) is shunted by a 5000-Ω load. Indicate the series resistor required in order to set the input firing voltage at 150 v?
2. What will be the values of the current in the load resistor and VR tube current immediately after the circuit of Problem 1 has fired?
3. Over what range of input voltage will the circuit of Problem 1 maintain 100 v across the load resistor?
4. If the 5000-Ω resistor across the tube of Problem 1 is varied, for what maximum and minimum values of R_L will the VR tube maintain 100 v across R_L? The input voltage remains at 155 v.
5. In the circuit of Fig. 4-2(a) $R_L = 1000\ \Omega$ and $E_{in} = 150$ v. Over what range can R_s be adjusted while maintaining 100 v across R_L? Use the tube characteristics of 4-2(b).
6. Referring to the circuit of Fig. 4-3(a), $R_L = 1000\ \Omega$. If the VR tube maintains 100 v from anode to cathode while the tube current varies from 1 to 8 mA, over what range of input voltage will the VR tube maintain this 100-v drop?
7. If R_L is 5 k in Fig. 4-3(a) and the VR tube regulates at 150 v, with 2 to 20 mA anode current, over what range may E_{in} be varied, while maintaining 150 v across the tube? What percentage change in the voltage across the resistor will occur as E_{in} varies over this range?
8. Calculate the frequency of operation for the circuit of Fig. 4-5(a) with the thyratron characteristics shown and $R_1 = 10\ \Omega$, $R_2 = 100\ \Omega$, and $C = 50\ \mu F$. The grid voltage is connected to a constant -4-v supply.
9. Referring to the circuit of Fig. 4-5 with tube characteristics as shown, if $R_2 = 150\ \Omega$ and $C = 100\ \mu F$, what value of grid voltage is necessary for a 60-Hz output frequency? What will the zero-to-peak value of the output voltage be?
10. If the grid voltage of the thyratron of Problem 9 is reduced to -2 v, what will the new peak output voltage and frequency be?
11. The circuit of Fig. 4-6 is to produce a peak-to-peak a-c voltage of 200 v across the 10-μF output capacitor. For an a-c input voltage of 60 Hz, calculate the values of R_1 and R_2.
12. For the circuit of Fig. 4-6(a), if $C = 20\ \mu F$, R_1 and $R_2 = 600\ \Omega$ each, and

the input is a 60-Hz signal, what will the peak-to-peak voltage across the capacitor be?

13. Calculate the firing angle of the thyratron tube represented in Fig. 4-9 when a 60-Hz sine wave with a peak voltage of 300 v is applied to the anode and when the grid voltage is: (a) -2 v, (b) -4 v, and (c) -6 v.
14. If a d-c voltage of -2 v and an a-c voltage of 2 v peak-to-peak were applied to the grid of the thyratron in Fig. 4-10(c), at what angle would the input voltage fire the tube? Let $\phi = 0$ initially and then $\phi = 30°$.
15. It is desired to fire the thyratron tube of Fig. 4-10(c) at 100°. A center-tapped transformer with a peak-output voltage of 4 v is available. Calculate the values of R and C required for the grid control when the anode of the thyratron is connected to a 60-Hz supply with a peak voltage of 500 v.
16. If R can vary from 0 to 5 k and $L = 6$ H for the phase-shift bridge of Fig. 4-11, over what angle (arc) can the voltage V_{BD} be adjusted? The input frequency is 60 Hz.
17. Calculate the firing angle for the thyratron represented by Fig. 4-10(c) when the a-c grid voltage is 2 v peak-to-peak and the d-c voltage is -3 v.
18. What value of d-c voltage will fire the thyratron of Fig. 4-10(c) at zero degrees when the a-c grid voltage is 2 v peak-to-peak?

Questions

1. Why will a small number of free electrons be present in a gas that is enclosed in a glass tube under low pressure?
2. When will the current in a gas-filled tube become self-maintaining?
3. Why can a gas-filled tube be used as a voltage-regulator?
4. Explain why the VR tube can be used as an error-amplifying device in a voltage-regulation circuit.
5. Why is the voltage at which a VR tube will fire higher than the voltage at which it will act as a regulator?
6. What is the function of the grid in a thyratron?
7. How will ambient temperature affect the operation of a thyratron?
8. What determines the value of the current through a thyratron after it has fired?
9. At what point in the operation of a thyratron does the grid become inoperative?
10. How is the current through a thyratron reduced to zero?
11. At what point in the operation of a thyratron will the grid regain control over the current?

12. What are the major differences between a-c and d-c thyratron circuits?
13. Describe the operation of an inverter circuit.
14. What determines the frequency of pulses in an inverter circuit?
15. How is the magnitude of the a-c voltage from the inverter circuit shown in Fig. 4-6(a) determined?
16. How is the frequency of the a-c output of the inverter circuit shown in Fig. 4-6(a) determined?
17. Describe the methods of grid control used in a-c and d-c thyratron circuits.
18. Explain why the firing angle is limited to 90° when a d-c voltage is used to control the firing angle of a thyratron.
19. How will the firing angle of the thyratron circuit shown in Fig. 4-8 be affected when the a-c input voltage is increased?
20. What are the advantages of a-c grid bias over d-c bias in thyratron circuits?
21. Using the grid locus curve of Fig. 4-10, explain how a shift in the phase of the grid voltage in relation to the anode-voltage changes the firing angle of the thyratron circuit.
22. Points *A* and *B*, on the phase-shifting network shown in Fig. E4-7, are interchanged, and point *A* is connected to the grid and point *B* to the cathode. Describe the operation of the circuit with this new connection.
23. Why is it more desirable to fire a thyratron with a vertical pulse than with a sine wave?

Controlled Rectifiers

5

The controlled rectifier provides a means of control over the operation of lighting circuits, welders, motors, and other industrial devices. The efficiency and ease of establishing remote or automatic control have made the controlled rectifier an indispensable addition to modern industrial-control circuits.

5-1 OPERATION OF CONTROLLED RECTIFIERS

A simple half-wave rectifying circuit that incorporates a vacuum-tube diode as the rectifying element is shown in Fig. 5-1(a). The average value of the load current, whose waveform is shown, can be expressed as $E_{m/\pi R}$.

In Fig. 5-1(b) a thyratron replaces the diode as the rectifying element. The conduction angle can now be controlled by varying E_g. It can be seen from the waveform of 5-1(b) that varying or controlling the conduction angle will cause corresponding variation or control of the average value of the load current.

A practical limitation of the thyratron tube is the maximum frequency at which it can be operated. The de-ionization time required for the grid to regain control over the tube limits its frequency of operation. A solid-state device that has replaced the thyratron in many applications is the *silicon-controlled rectifier* or SCR. This device can be operated at frequencies above 25 kc, does not require a separate cathode heater supply, and is more compact and efficient than the thyratron. Fig. 5-1(c) illustrates a half-wave rectifier with an SCR as the controlled rectifying-device. The SCR is used as the control element for all rectifying circuits discussed in this chapter. However, in each case similar circuits may incorporate a thyratron as the rectifying element.

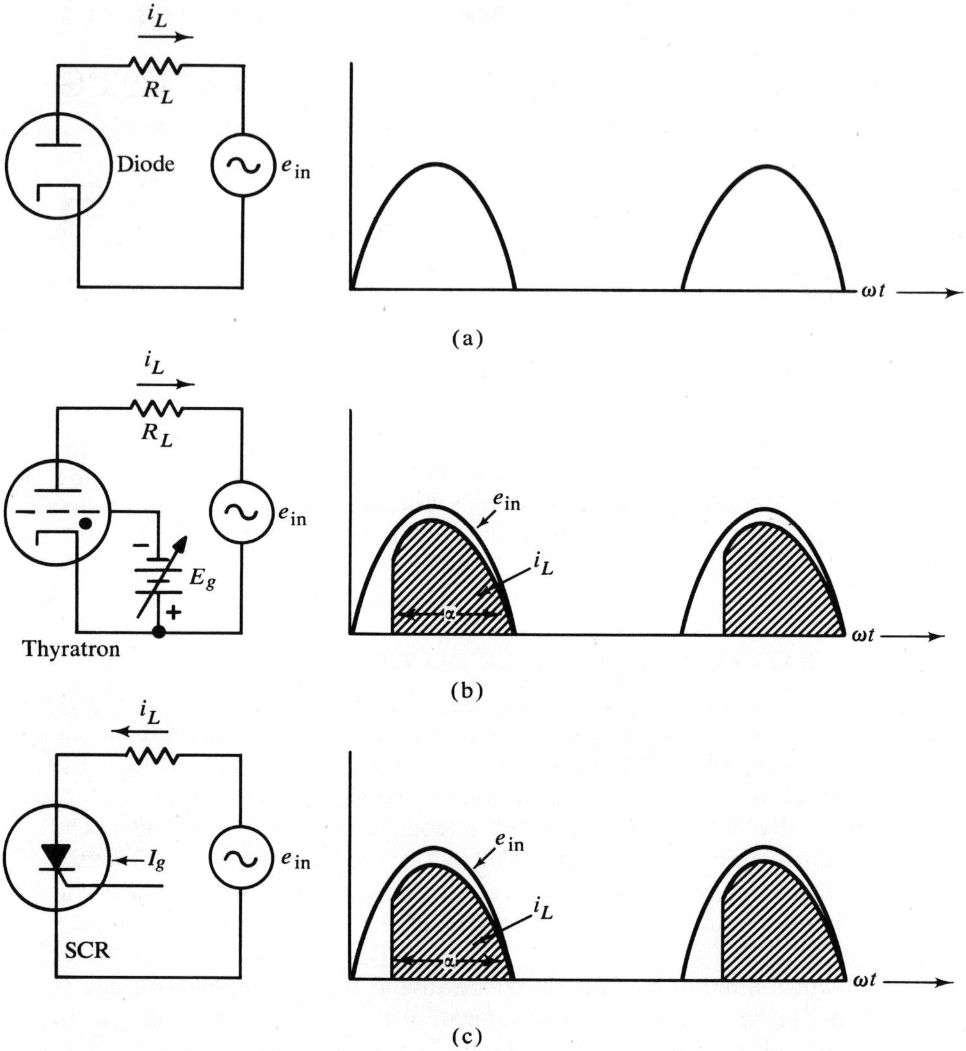

FIG. 5-1. (a) Half-wave rectifier circuit with vacuum-tube diode; (b) Half-wave rectifier circuit using thyratron tube, conduction angle equal to α; (c) Half-wave rectifier circuit using silicon-controlled rectifier (SCR), conduction angle equal to α.

5-2 THE SILICON-CONTROLLED RECTIFIER (SCR)

The SCR is similar in function to the thyratron and the SCR can be built with either the PNPN configuration shown in Fig. 5-2(a) or with the NPNP configuration.

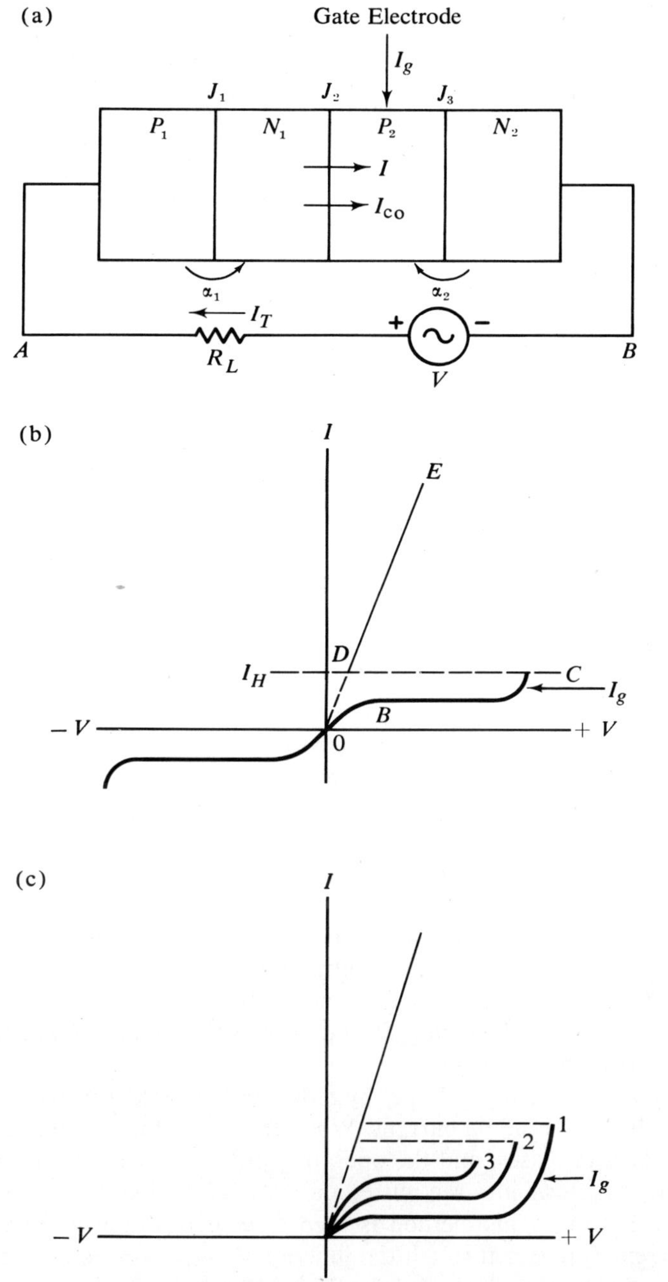

FIG. 5-2. (a) Silicon-controlled rectifier, type PNPN; (b) Load current vs source voltage for an SCR with fixed-gate current; (c) Load current vs source voltage for an SCR with various gate currents: $I_{g_3} \gg I_{g_2} \gg I_{g_1}$.

When point A of Fig. 5-2(a) is positive, the SCR is biased for conduction, and a forward bias is placed on junctions J_1 and J_3. Junctions J_1 and J_3 are forward-biased, and junction J_2 is reverse-biased. The potential across junction J_2, therefore, controls the current through the device. If I_g is zero, the total current I_T through junction J_2 will consist of $\alpha_1 I_T$ of holes from the P_1 region, $\alpha_2 I_T$ of electrons from the N_2 region, and a leakage current I_{co}. This may be expressed as

$$I_T = \alpha_1 I_T + \alpha_2 I_T + I_{co}$$

or

$$I_{co} = I_T(1 - \alpha_1 - \alpha_2)$$

and

$$I_T = \frac{I_{co}}{1 - (\alpha_1 + \alpha_2)} \tag{5-1}$$

The leakage current I_{co} is very small in silicon, and at low-emitter currents the values of α_1 and α_2 are also small. Therefore, the value of I_T is very low when I_g is zero. However, if the emitter current is increased by raising the value of V or increasing I_g, α_1 and α_2 will increase rapidly. If the value of $(\alpha_1 + \alpha_2)$ varies in Eq. 5-1, the current I_T will also vary; as $(\alpha_1 + \alpha_2)$ approaches unity, the current I_T will become very large.

At low currents in the region 0 to B of the curve in Fig. 5-2(b), the value of $(\alpha_1 + \alpha_2)$ is <1, and the junction J_2 acts as a reverse-biased PN junction. In the region B to C, $(\alpha_1 + \alpha_2)$ is increasing, and at point C, $(\alpha_1 + \alpha_2) \cong 1$. C is the breakover or triggering point. (The SCR may be triggered by increasing either V or I_g.) At point C the potential at junction J_2 is reduced, and the SCR behaves as if it were a single PN junction biased in the forward direction. The device operates in the region D to E of the volt-ampere curve with a high current and low voltage drop.

Conduction can be stopped only by reducing the current below the critical value I_H, called the holding current. When the current is reduced below this value, conduction ceases, and the operating point returns to point 0.

If a current I_g flows into section P_2 and across junction J_3, α_2 will increase independently of α_1. This method is used to control the firing of the SCR. As the current I_g passes through the junction J_3, α_2 increases. The condition $(\alpha_1 + \alpha_2) = 1$ is approached, and the SCR is fired.

The operation of the SCR can also be analyzed from the standpoint of the potential levels within the PN sections. In Fig. 5-3 the various potential levels are shown. The junctions J_1 and J_3 are forward-biased, and junction J_2 is reverse-biased when the source has the polarity shown. The potential barrier at J_2 can be lowered by means of gate current I_g through the P_2 section of the SCR. After the barrier at J_2 is lowered, current flows through the device from section P_1 to section N_2 and SCR conducts in the forward direction.

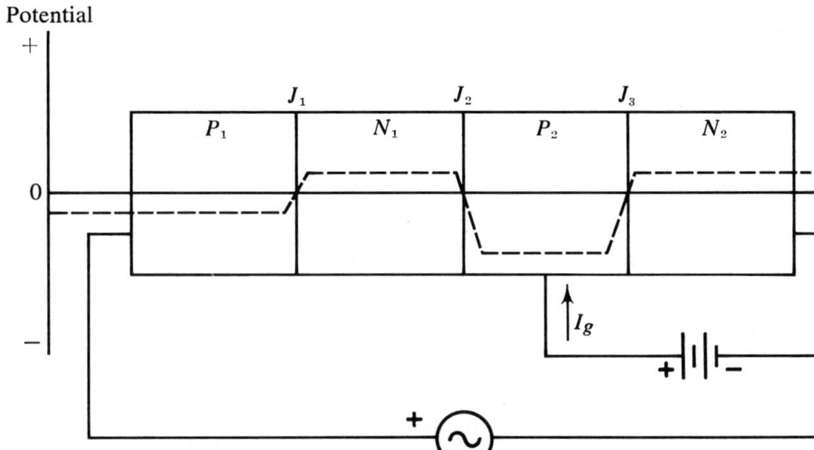

FIG. 5-3. Potential levels established in SCR by external biasing.

If I_g is reduced to zero, after conduction starts, the forward current flow causes the barrier at J_2 to remain lowered, and the SCR continues to conduct.

The curves of Fig. 5-2(c) illustrate how the triggering current varies with supply voltage. When triggering current is high, the SCR breaks or fires at low values of the supply voltage. In normal operation the peak a-c supply voltage is well below the breakover value for $I_g = 0$, permitting the SCR to be triggered at the desired firing angle by a triggering current applied to the gate-electrode. Once the device is in the conduction state, the gate current no longer has control over the load current. The only method of stopping conduction is to reduce the SCR load current below the holding value I_H. As in thyratron circuits, conduction is halted by a reduction in or reversal of the supply voltage.

The SCR is triggered by a current pulse, the thyratron by a voltage pulse. The SCR, therefore, requires triggering circuits with lower internal impedance.

5-3 TRIGGERING THE SILICON-CONTROLLED RECTIFIER

Various firing circuits have been developed to trigger the SCR. In Fig. 5-4 a simple SCR firing circuit is shown which draws the required gate current from the a-c supply. When the switch is closed, gate current flows and the SCR conducts when its anode voltage is of the required positive value. When the SCR conducts, the anode-to-cathode voltage drops to a very low value,

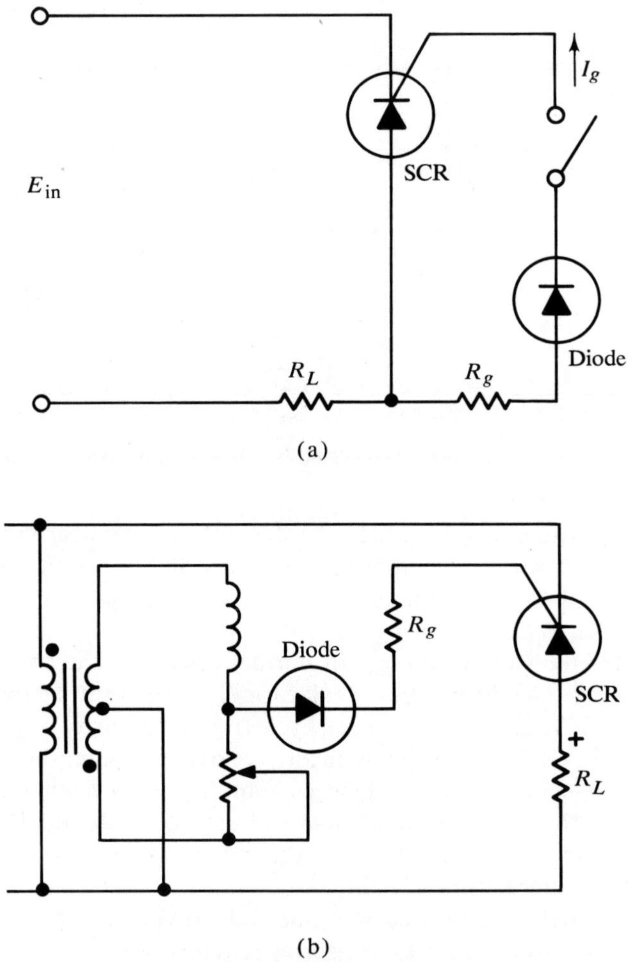

FIG. 5-4. (a) SCR circuit with diode trigger; (b) SCR using phase-shift bridge in trigger circuit.

and the gate current is reduced to approximately zero. R_g limits the gate current to a safe value before the voltage is reduced. The diode prevents an inverse current from flowing in the gate during the negative half-cycle. Conduction ceases at the end of the conducting half-cycle when the anode voltage goes negative. This circuit can control the firing angle only from zero to 90°. The phase-shift bridge must be employed if control is desired over a wider range.

In Fig. 5-4(b) a conventional phase-shift bridge is used to control the gate current. A diode has been added to protect the SCR during the inverse half-cycle.

5-4 HALF-WAVE CONTROLLED RECTIFICATION

The silicon-controlled rectifier is used in half-wave rectifier circuits in which control of the average d-c output is desired. By varying the angle at which the device is fired, the period of current conduction is controlled and, therefore, the d-c output is controlled. Figure 5-5 shows the basic half-wave control-rectifier circuit as well as its current and voltage waveforms. The gate current is such that after the supply voltage has rotated through the angle ϕ, reaching the value E_f, the SCR is triggered, and the load current I_L rises to the value I_f. The current then follows the supply voltage until the supply voltage reverses at π. At that time the SCR is cut off until the next positive cycle has progressed through ϕ degrees.

The average load current and voltage are found by integrating the waveforms for one cycle and then dividing by the period of one cycle.

Assuming zero voltage drop across the SCR during conduction, the average voltage across the load is expressed as

$$E_{\text{d-c}} = \frac{1}{2\pi}\left[\int_0^{\phi} 0(d\omega t) + \int_{\phi}^{\pi} E_{\text{m}} \sin \omega t(d\omega t) + \int_{\pi}^{2\pi} 0(d\omega t)\right]$$

then

$$E_{\text{d-c}} = \frac{1}{2\pi}\int_{\phi}^{\pi} E_{\text{m}} \sin \omega t(d\omega t)$$

and

$$E_{\text{d-c}} = \frac{E_{\text{m}}}{2\pi}(1 + \cos \phi) \tag{5-2}$$

$I_{\text{d-c}}$ is found by the following equation.

$$I_{\text{d-c}} = \frac{E_{\text{m}}}{2\pi R}(1 + \cos \phi) \tag{5-3}$$

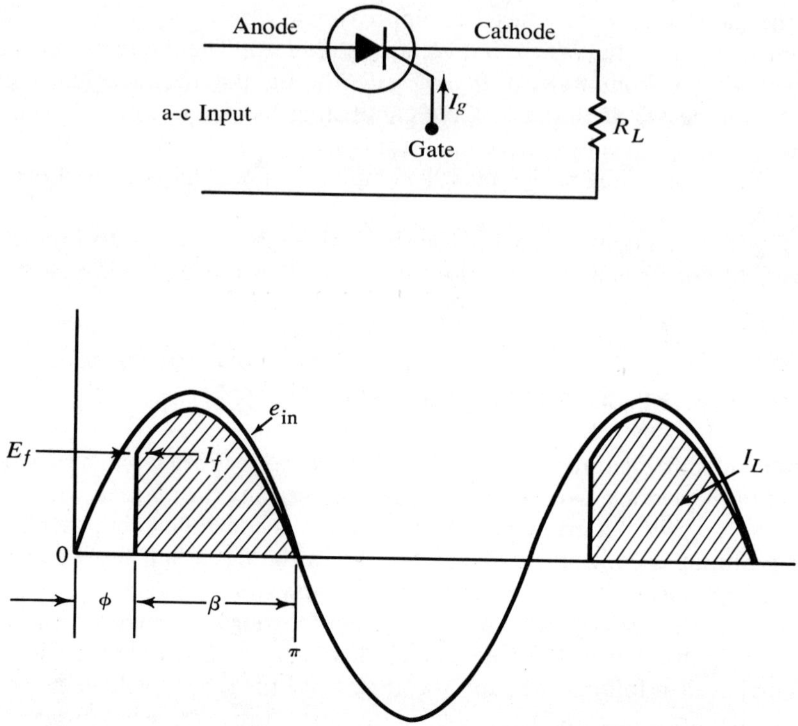

FIG. 5-5. (a) Half-wave SCR rectifier circuit; (b) Input-voltage and load-voltage waveforms.

The average voltage across the load is, therefore, a function of the firing angle and maximum supply voltage. An important quantity is the value of the maximum peak current, which is found after assuming the drop across the SCR to be zero. Then

$$I_m = \frac{E_m}{R} \tag{5-4}$$

The average power in the circuit is defined as $P_{av} = I_{rms}{}^2 R$.

$$I_{rms} = \frac{E_m}{2\pi R}\left[\int_{\phi}^{\pi} (\sin \omega t)^2 \, d\omega t\right]^{1/2}$$

and

$$P_{av} = \frac{(E_m)^2}{4\pi R}\left(\beta + \frac{1}{2}\sin 2\phi\right) \tag{5-5}$$

where β is the conduction angle.

Example 5-1

The half-wave SCR rectifier of Fig. E5-1 is being fired from a phase-shift bridge with the values of R and C as shown. Assume that the SCR fires at the moment the voltage V_{BD} goes positive. Find the:

1. firing angle ϕ
2. conduction angle β
3. average d-c load current

1. Calculate the firing angle as follows.

The vector diagram for the phase-shift bridge can be drawn as shown in Fig. E5-1(b). Since an RC circuit is being used the current I will lead the voltage V_{CA}. The voltage V_R will be in phase with the current, and the voltage V_C will lag the current by 90°. The impedance of the phase-shifting network must now be calculated.

$$Z = 5000 + \frac{1}{J\,377(0.4)10^{-6}}$$

or

$$Z = 5000 - J\,6600$$

Rewriting

$$Z = 8350\angle{-53^\circ}$$

The current in the phase-shifting network is next calculated.

$$I_B = \frac{120}{8350\angle{-53^\circ}} = 0.0144\text{ A}\angle 53^\circ$$

$$V_R = 72\angle 53^\circ$$

$$V_C = (0.0144\angle 53^\circ)(6600\angle{-90^\circ})$$

$$V_C = 96\angle{-37^\circ}$$

Redrawing the vector diagram of Fig. E5-1(b) as shown in Fig. E5-1(c), the SCR will be triggered by the voltage V_{BD}. The angle ϕ is the firing angle. Assuming that the input voltage is in phase with the secondary voltage V_{CA} of the transformer, and, assuming further that the SCR fires at the instant the gate voltage goes plus, solve for ϕ as follows.

Since a center-tapped transformer is being used, $V_{BC} = V_{BD}$ and $\angle BCD = \angle CDB$,

and

$$\phi = 180^\circ - 2(53^\circ)$$

or

$$\phi = 74^\circ$$

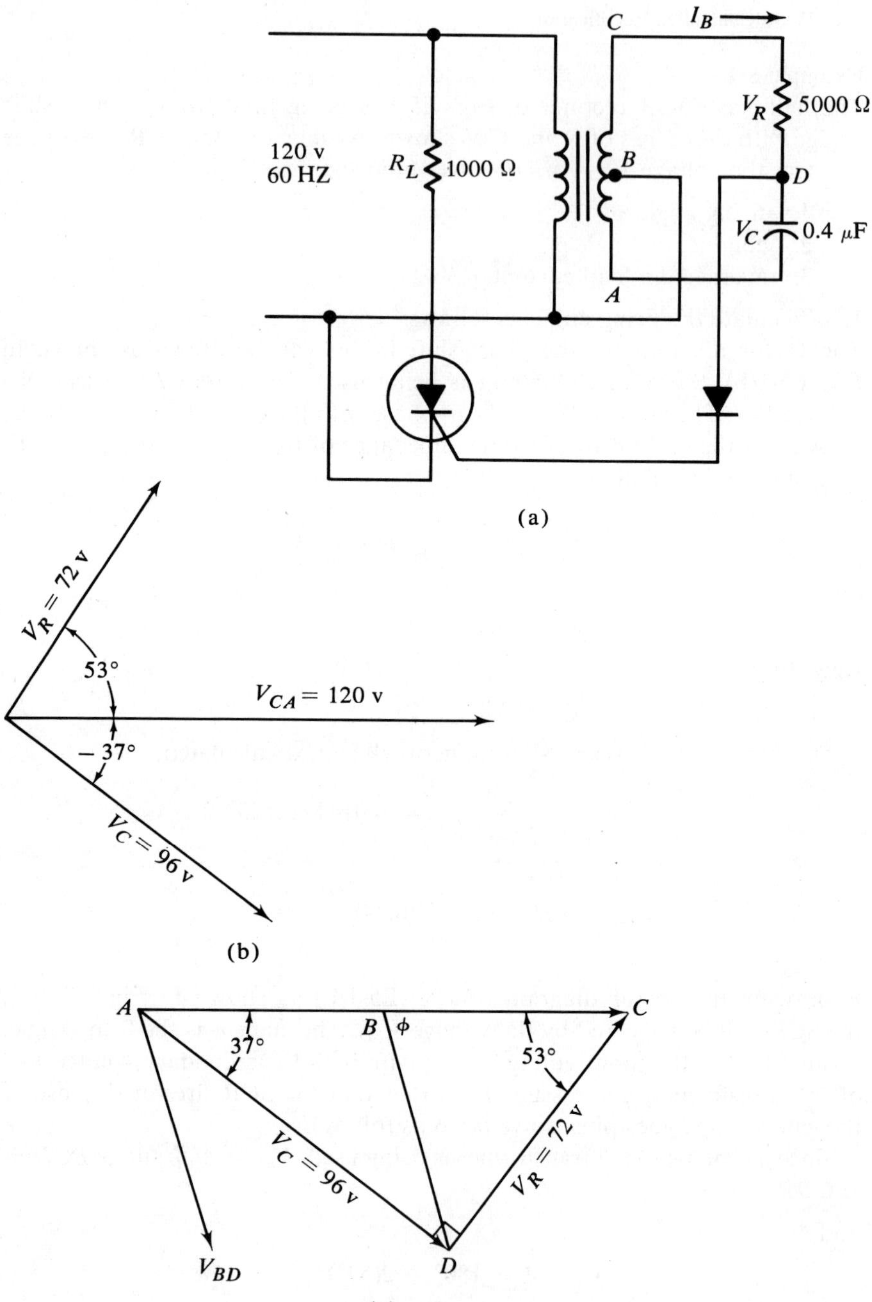

FIG. E5-1. Figure for Example 1.

These results can be confirmed by Eq. 4-2.

2. The conduction angle is (180 − 74)° or 106°.

3. Calculate I_{av} as follows. Since the firing angle is known, Eq. 5-3 is used to find the average current.

$$I_{d\text{-}c} = \frac{E_m}{2\pi R}(1 + \cos\phi)$$

$$I_{d\text{-}c} = \frac{170}{6280}(1 + 0.275)$$

$$I_{d\text{-}c} = 34.5 \text{ mA}$$

5-5 FULL-WAVE CONTROLLED RECTIFIER WITH RESISTANCE LOAD

The full-wave controlled-rectifier circuit is widely used in motor-control applications. A typical circuit, shown in Fig. 5-6, incorporates a plot of supply voltage and load voltage as a function of time.

The average load voltage is again a function of the firing angle and is exactly twice the value of that of a half-wave circuit. Therefore, the expression for average voltage and current may be written as

$$E_{d\text{-}c} = \frac{E_m}{\pi}(1 + \cos\phi) \tag{5-6}$$

and

$$I_{d\text{-}c} = \frac{E_m}{\pi R}(1 + \cos\phi) \tag{5-7}$$

where ϕ is the firing angle.

The instantaneous peak current in the full-wave controlled rectifier is the same as in the half-wave device.

$$I_m = \frac{E_m}{R}$$

The total power delivered to the load is double that in the half-wave device, or

$$P_{av} = \frac{(E_m)^2}{2\pi R}\left(\beta + \frac{1}{2}\sin 2\phi\right) \tag{5-8}$$

where β is the angle through which the SCR conducts.

Example 5-2 For the circuit shown in Fig. 5-6, $E_m = 200$ v; $R_L = 100\ \Omega$; and $\phi = 30°$. At the output of the rectifier circuit find E_{av} and P_{av}.

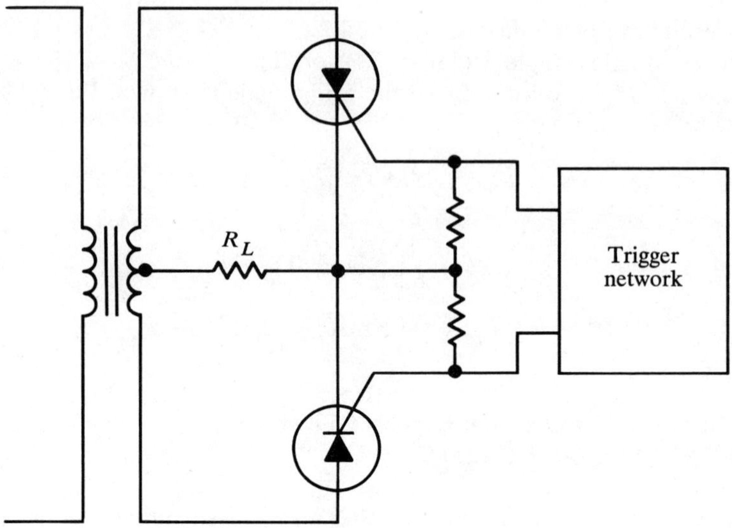

(a)

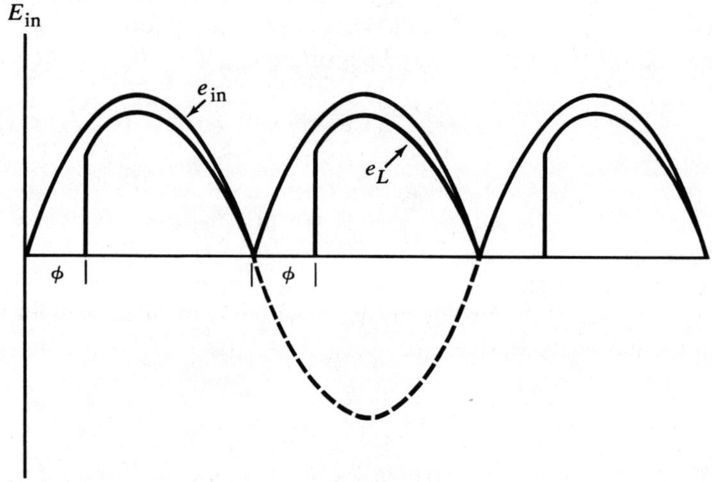

(b)

FIG. 5-6. (a) Full-wave SCR rectifier circuit; (b) Input-voltage and load-voltage waveforms.

From Eqs. 5-6 and 5-8

$$E_{av} = \frac{200}{\pi}(1 + \cos 30°)$$

$$E_{av} = 119 \text{ v}$$

$$P_{av} = \frac{(200)^2}{2\pi R}\left(\frac{5\pi}{6} + \frac{1}{2}\sin 60°\right)$$

$$P_{av} = 63.7(2.62 + 0.433)$$

$$P_{av} = 194 \text{ W}$$

5-6 FULL-WAVE CONTROLLED RECTIFIER WITH *RC* LOAD

When a rectifier circuit supplies power to an *RC* load, the diodes conduct only during that portion of the cycle when the input voltage exceeds the voltage on the capacitor. This type of circuit is discussed in Chapter 1. When the diodes are replaced by SCR's, as in Fig. 5-7, the firing angle and the charge on the capacitor determine the period during which the rectifier conducts. The rectifier of Fig. 5-7(b) would normally conduct when the input voltage rises just to a point barely above the value e_C. At this point the input waveform has passed through the angle α. As discussed in Chapter 1, the rectifier current must rise to a high value in order to charge the capacitor to the peak line-voltage in the short time available. However, if triggering of the SCR is delayed until the voltage vector has passed through the angle ϕ, the conduction period is shorter and peak-current value must be larger. The voltage across the SCR before firing is equal to the difference between the voltage on the capacitor and the line voltage. Since this voltage is much lower than the input voltage, a higher current is required to trigger the SCR when it is supplying an *RC* load. The d-c output voltage is determined in the same manner as that used in Chapter 1 for a rectifier with an *RC* load. If the firing angle is between zero and 90°,

$$E_{d\text{-}c} = E_m - \frac{\Delta E}{2} \tag{5-9}$$

where

$$\Delta E = \frac{E_{d\text{-}c}}{2fRC} \tag{5-10}$$

therefore

$$E_{d\text{-}c} = E_m\left(\frac{4fRC}{4fRC + 1}\right) \tag{5-11}$$

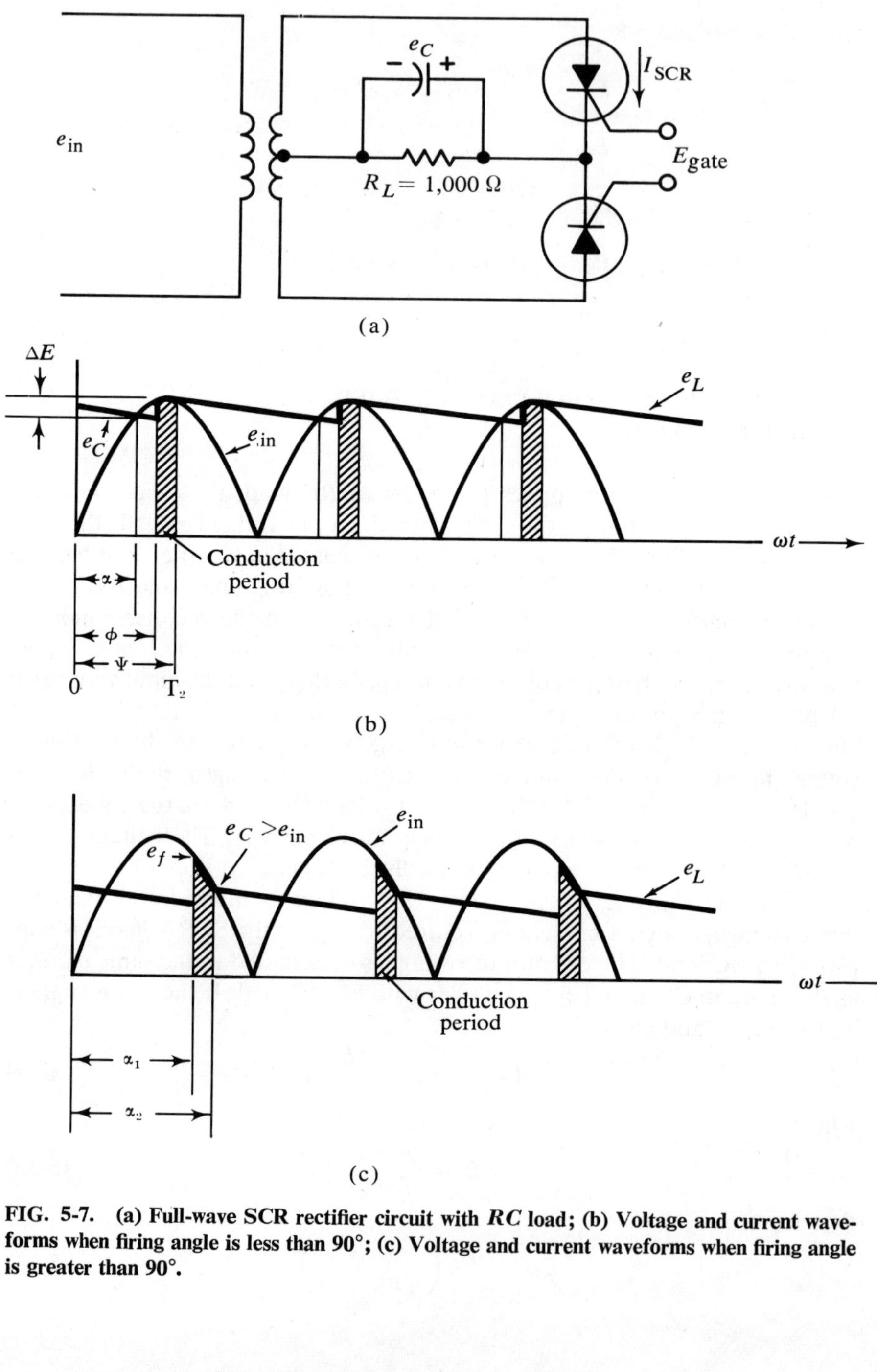

FIG. 5-7. **(a) Full-wave SCR rectifier circuit with *RC* load; (b) Voltage and current waveforms when firing angle is less than 90°; (c) Voltage and current waveforms when firing angle is greater than 90°.**

The ripple factor can be expressed as

$$\gamma = \frac{0.29}{2fRC} \tag{5-12}$$

From the preceding relationships it is seen that as the RC time constant increases, $E_{d\text{-}c}$ approaches E_m and the a-c ripple decreases.

Example 5-3 For the circuit shown in Fig. 5-7 the maximum load voltage is 170 v. If $R_L = 1000$ and $C = 50\ \mu$F, find the average SCR current during conduction when the firing angle ϕ is equal to: (a) zero and (b) 80°.

1. When $\phi =$ zero. The SCR will conduct when $e_{\text{in}} = e_C$. Treating the circuit as a full-wave bridge, from Eq. 1-48

$$E_{av} = E_m\left(\frac{4fRC}{4fRC + 1}\right)$$

If

$$f = 60 \text{ Hz}$$

$$E_{av} = \frac{12}{13}(170) = 157 \text{ v}$$

From Eq. 1-47

$$\Delta E = \frac{E_{av}}{2fRC}$$

$$\Delta E = 26 \text{ v}$$

From Eq. 1-36

$$\alpha = \sin^{-1}\left(\frac{E_m - \Delta E}{E_m}\right)$$

$$\alpha = 57.5°$$

From Eq. 1-26

$$\psi = \tan^{-1}(-\omega RC)$$

$$\psi = 93°$$

The average SCR current during conduction is obtained from the average load current and the conduction time.

$$I_{av} = E_{av}/R$$

$$I_{av} = 0.157 \text{ A}$$

The capacitor must supply this current when the SCR is not conducting, or for approximately one-half cycle.

The charge, Q, supplied by the capacitor is

$$Q = 0.157 \text{ A}(\tfrac{1}{120} \text{ sec})$$

from which $Q = 0.00131$ C. This charge must be supplied while the SCR is conducting. During conduction the SCR also supplies current to the load resistor.

$$I_{SCR} = JQ/\text{conduction time} + E_{av}/R$$

Before the SCR current can be calculated, the conduction time is found as follows:

From previous calculation

$$\alpha = 57.5°$$

and

$$T_1 = 57.5° \times \frac{1}{360} \times \frac{1\text{ sec}}{60} = 0.0026\text{ sec}$$

As previously calculated,

$$\psi = 93°$$

and

$$T_2 = 0.0043\text{ sec}$$

The conduction time is equal to $T_2 - T_1$, or 0.0017 sec. The average SCR current is next calculated.

$$I_{SCR} = \text{J}\frac{0.00131\text{ C}}{0.0017\text{ sec}} + \frac{157}{1000}$$

or

$$I_{SCR} = \text{j}0.77 + 0.157 = 0.78\text{ A}$$

If conduction is delayed until $\phi = 80°$, the conduction time will decrease. The new conduction time is calculated as follows:

$$\phi = 80°$$

Therefore

$$T_1 = 0.0037\text{ sec}$$

$$T_2 = \text{remains at } 0.0043\text{ sec}$$

and

$$T_2 - T_1 = 0.0006\text{ sec}$$

$$E_{av} = \frac{E_m}{\pi}(1 + \cos 80) \qquad \text{(Eq. 5-6)}$$

$$E_{av} = 63\text{ v}$$

and

$$I_{av} = 0.063\text{ A}$$

The charge Q, supplied by the capacitor, is

$$Q = 0.063 \text{ A} \left(\tfrac{1}{120}\right) \text{sec}$$

$$Q = 0.00052 \text{ C}$$

The average SCR current during conduction is then

$$I_{\text{SCR}} = J\frac{0.00052}{0.0006} + 0.063$$

$$I_{\text{SCR}} = J0.87 + 0.063 = 0.875 \text{ A}$$

5-7 CONTROLLED-RECTIFIER CIRCUITS WITH *RL* LOAD

One of the most common types of loads encountered with SCR circuits is the *RL* load. Generator fields, motor controllers, and arc welders are examples of this type of load.

As an aid in analysis of the operation of the full-wave SCR circuit with *RL* load and zero firing angle, the waveform for such a circuit is displayed in Fig. 5-8(b). When the supply voltage starts at $T = 0$ and the voltage at point *A* is swinging positive, the current through SCR 1, produces a sinusoidal voltage across the *RL* load. The Kirchhoff voltage equation for the conducting loop is

$$e_{\text{in}} = iR + L\frac{di}{dt}$$

The current, dependent upon both the time constant of the load and the rate of change of source voltage, follows the equation

$$i = \frac{E_{\text{m}} \sin \phi}{Z} e^{-Rt/L} + \frac{E_{\text{m}} \sin (\omega t - \phi)}{Z}$$

for the first half-cycle. At the end of 180° the supply voltage becomes zero, but the inductance will maintain current in SCR 1. Point *E* of the inductance will be positive with respect to point *D*, owing to the decay of the current through *L*.

As the voltage at point *B* swings positive on the next half-cycle, the induced voltage e_L is added to the input voltage, and SCR 2 will conduct immediately. The induced voltage aids in the starting of conduction through SCR 2. When SCR 2 begins to conduct, the current in the inductance increases and the induced voltage e_L reverses polarity, making point *D* more positive than *A*.

After point *D* becomes positive, SCR 1 is cut off and SCR 2 conducts because point *B* is more positive than *D*. The current has shifted from SCR 1

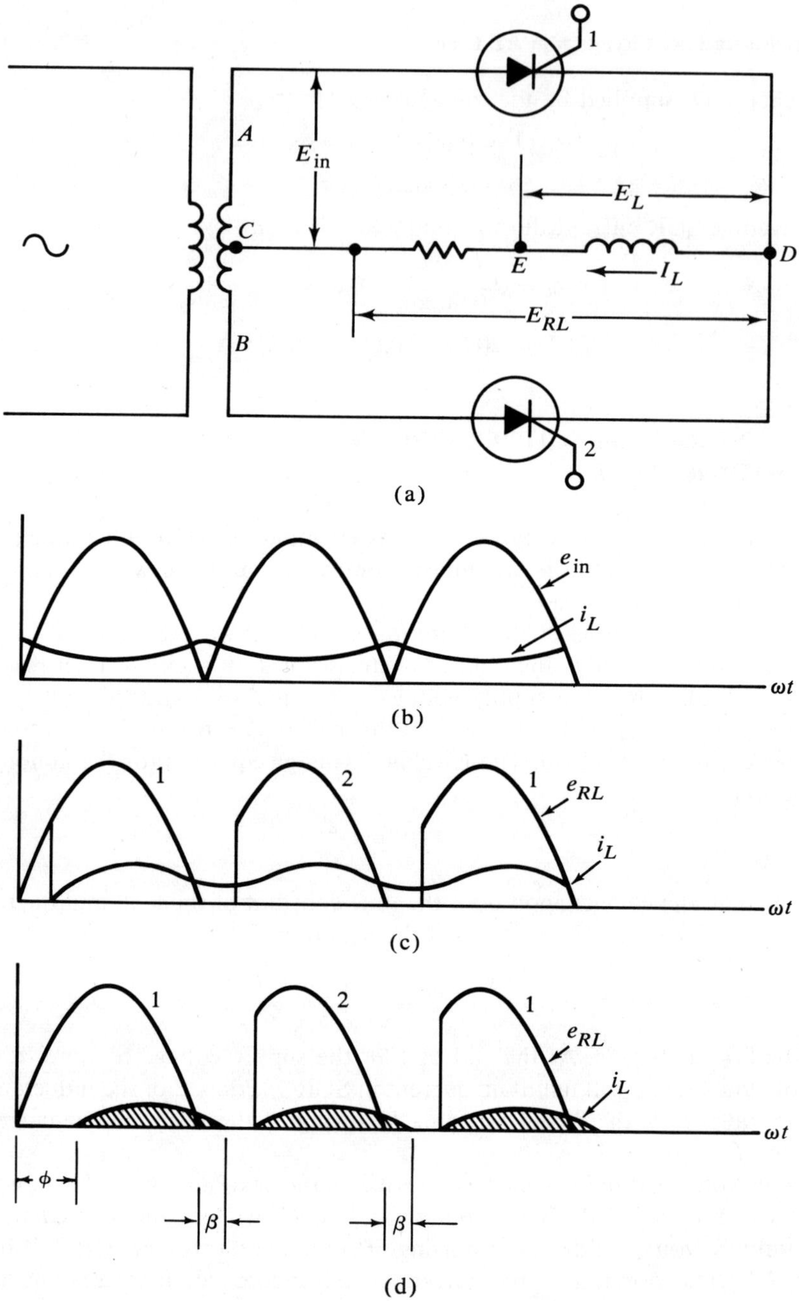

FIG. 5-8. (a) Full-wave SCR rectifier circuit with *RL* load; (b) Voltage and current waveforms when firing angle is equal to zero; (c) SCR full-wave rectifier, mode 1 operation; (d) SCR full-wave rectifier, mode 2 operation.

to SCR 2 by a process known as commutation, which takes place very half-cycle. The line voltage passes through zero, but the load current is never zero.

When the firing angle is not zero, the current and voltage exhibit the waveforms shown in part (c) or (d) of Fig. 5-8. Two possible conditions can exist: mode 1 operation and mode 2 operation. In mode 1 operation, SCR 1 conducts when SCR 2 is fired; in mode 2 operation, SCR 1 is cut off when SCR 2 fires.

Figure 5-8(c) shows the waveforms for mode 1 operation. The triggering angle is such that SCR 1 is still conducting when SCR 2 fires. Conduction through SCR 2 takes place immediately, and SCR 1 is still conducting when SCR 2 starts to conduct. Commutation takes place at the angle ϕ, rather than at zero, as in the case of the zero firing angle shown by the waveforms of Fig. 5-8(b).

Figure 5-8(d) shows the waveforms for mode 2 operation. Here the triggering occurs after the current in the conducting SCR has reached zero. A period during which neither diode is conducting exists every half-cycle.

Mode 1 is by far the most commonly used type of operation. Assuming that the drop across each SCR is small enough to be ignored, for mode 1 operation it can be stated:

$$E_{\text{d-c}} = \frac{2E_{\text{m}}}{\pi} \cos \phi \tag{5-13}$$

and

$$I_{\text{d-c}} = \frac{2E_{\text{m}}}{\pi R} \cos \phi \tag{5-14}$$

The average power in the load can be expressed as approximately

$$P_{\text{av}} \cong \frac{4(E_{\text{m}})^2}{\pi^2 R} (\cos \phi)^2 \tag{5-15}$$

5-8 CONTROLLED THREE-PHASE RECTIFIERS

The three-phase rectifier is another widely used circuit that employs SCR devices to control the average d-c output. A full-wave SCR bridge is shown in Fig. 5-9. In order for an SCR to conduct, two conditions must be met. The SCR must be biased in the forward direction, and the gates must also be properly biased. This limits the conduction of each SCR to 120° in a full-wave three-phase bridge. Control of the SCR must, therefore, take place in the region between 30° and 150° for each phase voltage. The firing angle must be greater than 30° if control of the d-c power is to be obtained. If ϕ is less

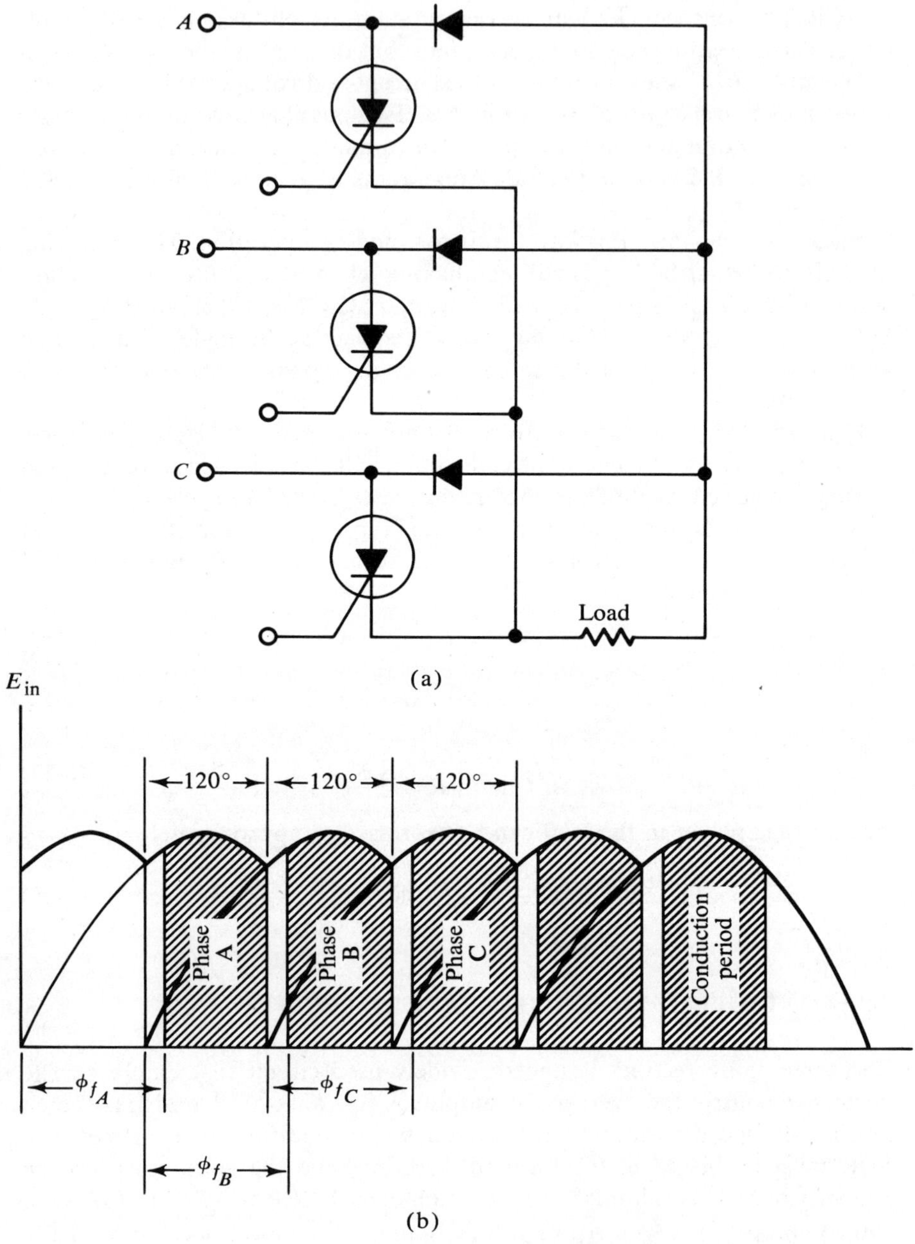

FIG. 5-9. (a) Three-phase, full-wave SCR bridge rectifier; (b) Voltage and current waveforms.

than 30°, the diodes conduct during the interval 30° to 150°, and the average d-c power in the load is the same as that obtained from a standard three-phase bridge without an SCR. As ϕ is increased beyond 30°, conduction is prevented during part of the normal conduction period, and control over the average d-c power is obtained.

The average load voltage can be expressed as

$$E_{\text{d-c}} = \frac{3E_m}{2\pi}\int_{\phi_f}^{2\pi/3} \sin \omega t(d\omega t)$$

When

$$30° < \phi_f < 150°$$

then

$$E_{\text{d-c}} = 0.478\, E_m\left(\cos \phi_f \cos \frac{2\pi}{3}\right) \qquad \textbf{(5-16)}$$

5-9 BACK-TO-BACK CIRCUIT WITH RESISTANCE LOAD

The back-to-back SCR circuit shown in Fig. 5-10 is commonly used to control power required by electric furnaces, lighting systems, and other types of applications. In the type of circuit shown in Fig. 5-10, power to the load is controlled by the restriction of current flow to less than a full cycle. Since the current in the load reverses direction, a-c power is delivered.

With a sinusoidal input voltage SCR 1 begins to conduct at the angle ϕ_1. Conduction through SCR 1 continues until the input voltage is zero, at which time SCR 1 is cut off. SCR 2 then conducts when the negative half-cycle passes through ϕ_2°. We have assumed that $\phi_1 = \phi_2$, which is usually true as a result of the manner in which these circuits are designed. However, this condition is not absolutely necessary, and cases are found in which $\phi_1 \neq \phi_2$, for instance, SCR 1 is replaced by a diode and only SCR 2 is controlled. When $\phi_1 = \phi_2$ and the drops across each SCR is assumed to be zero, the load voltage may be expressed as

$$E_{\text{rms}} = \frac{E_m}{\sqrt{2}}\sqrt{1 - \frac{1}{\pi}\left(\frac{\sin 2\phi}{2} + \phi\right)} \qquad \textbf{(5-17)}$$

and the load power as

$$P_{\text{av}} = \frac{E_m{}^2}{2R}\left[1 - \frac{1}{\pi}\left(\frac{\sin 2\phi}{2} + \phi\right)\right] \qquad \textbf{(5-18)}$$

Example 5-4 What will be the effective voltage and power in the load for the circuit in Fig. 5-10(a) if the firing angle is 60°, $E_{\text{in}} = 220$ v and $R_L = 200\,\Omega$?

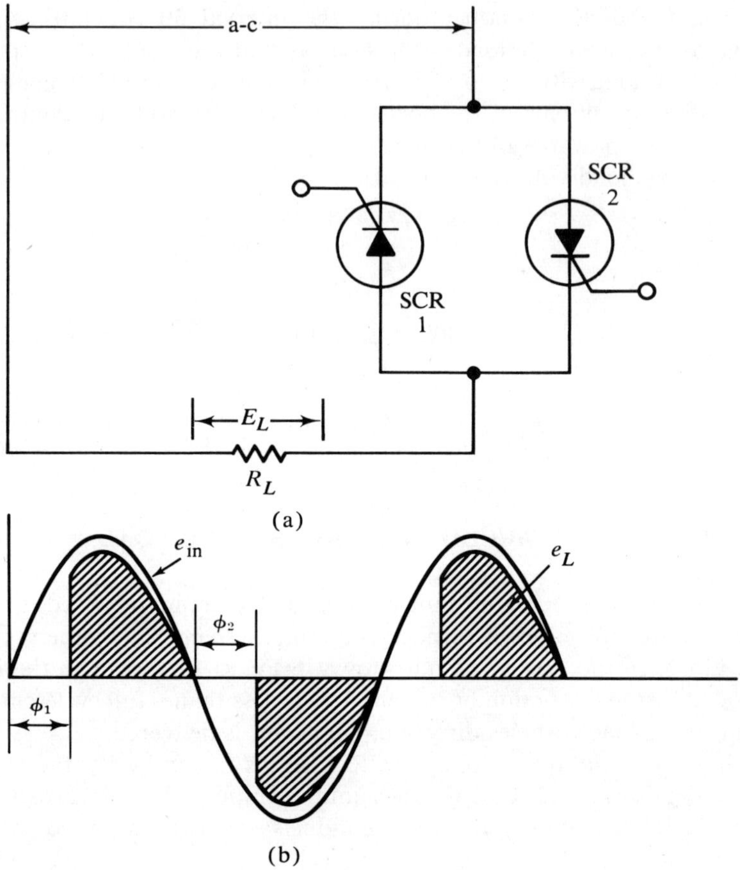

FIG. 5-10. (a) Back-to-back SCR circuit; (b) Voltage and current waveforms.

From Eq. 5-17

$$E_{\text{rms}} = \frac{220\sqrt{2}}{\sqrt{2}}\left\{\sqrt{1 - \frac{1}{\pi}\left[\frac{1}{2}\sin(120) + \frac{\pi}{3}\right]}\right\}$$

$$E_{\text{rms}} = 220\sqrt{1 - (0.138 + 0.33)}$$

$$E_{\text{rms}} = 220\sqrt{0.532}$$

$$E_{\text{rms}} = 159 \text{ v}$$

From Eq. 5-18

$$P_{\text{av}} = \frac{(220\sqrt{2})^2}{2 \times 200}(0.532)$$

$$P_{\text{av}} = 128 \text{ W}$$

5-10 BACK-TO-BACK CIRCUIT WITH *RL* LOAD

More frequently found in combination with the resistance-inductance load is the back-to-back circuit, a type of which is presented in Fig. 5-11, together with the waveforms encountered.

Two modes of operation are possible for this circuit: modes 1 and 2. In mode 1, illustrated by the waveforms in Fig. 5-11(b), SCR 1 is still conducting when SCR 2 is triggered. In mode 2, illustrated by the waveforms in Fig. 5-11(c), SCR 1 has ceased to conduct when SCR 2 is triggered.

In mode 1 SCR 2 fires while SCR 1 is conducting. The induced load voltage keeps point B negative with respect to A, and SCR 2 will not immediately conduct when fired. SCR 2 conducts only after the load current has reached zero. At that point the load current reverses. It is interesting to note here, that the load current controls the conduction of the next SCR and not the gate current. The angle α is a function of L/R, which is the time constant of the load. If the firing angle ϕ is less than the time constant angle α, the point of load-current-reversal is determined by the angle α and not by the firing angle ϕ. Figure 5-11(b) shows the waveforms of current and voltage when $\alpha > \phi$.

When the firing angle is larger than the time constant angle, the current in the conducting SCR will be zero at the time when the next SCR is triggered and operation is in mode 2. From the waveforms of current and voltage for mode 2 given in Fig. 5-11(c), the load current is seen to be zero when SCR 1 is triggered. The line voltage appears across the load instantaneously, and the load current begins to rise at a rate determined by the time constant L/R. When the source voltage passes it peak value and begins to fall, the polarity of the induced load voltage reverses. From point 0 to point A on the curve of Fig. 5-11(c) SCR 1 conducts, because the decay of current in the inductance keeps point B minus with respect to point C, even though the source voltage has reversed. The angle of conduction at the beginning of the next half-cycle is shown as the angle α. When the negative half-cycle has progressed through $\beta°$, SCR 2 is fired and the cycle repeats.

5-11 LOAD-MATCHING BY USE OF TRANSFORMERS

It has been shown that the back-to-back circuit delivers an a-c current to the load. Such current may be continuous, as in mode 1 operation, or pulsating, as in mode 2 operation. Both the continuous and pulsating currents may be used in conjunction with transformers in order to match a source of power

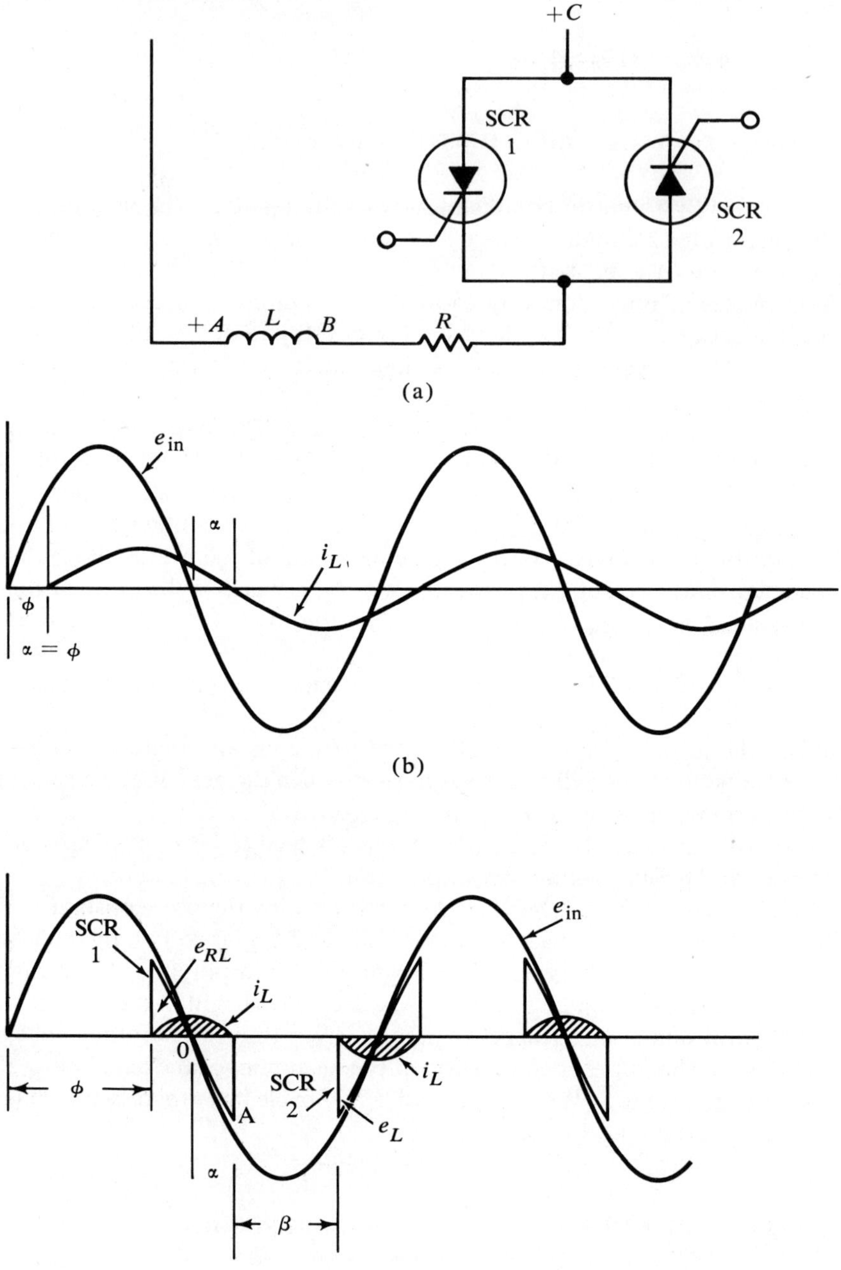

FIG. 5-11. (a) Back-to-back SCR circuit with *RL* load; (b) Voltage and current waveforms in mode 1 operation; (c) Voltage and current waveforms in mode 2 operation.

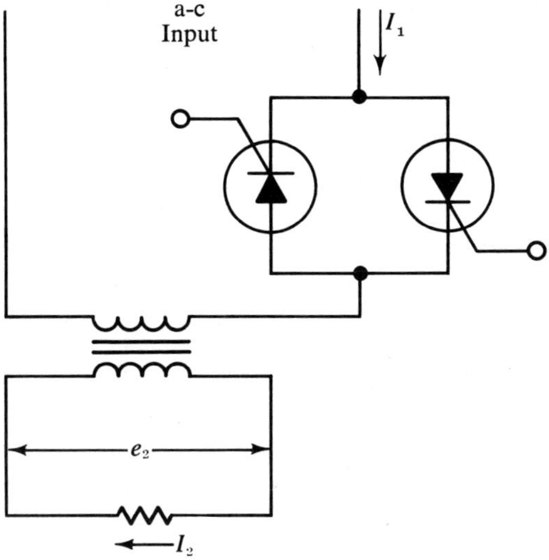

(a)

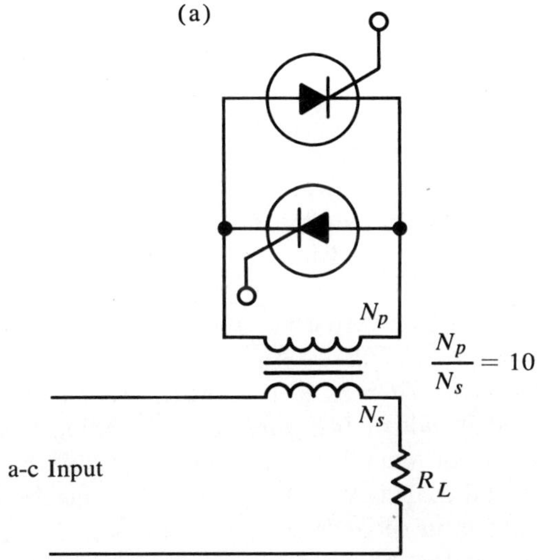

(b)

FIG. 5-12. (a) Transformer used for matching SCR rating to load current; (b) Series inductance controlled by varying the firing angle of SCR.

to a load. Figure 5-12(a) shows a circuit in which two SCR's, each capable of passing 20 A, are matched to a welder requiring 200 A.

Basic transformer theory states that $e_1/e_2 = i_1/i_2$. Therefore, if it is desired to deliver 200 A to a welder at a low voltage, two rectifiers, each capable of carrying 20 A, and a matching transformer with a 10-to-1 turns ratio may be used. Alternatively, a 100-to-1 matching transformer and devices rated at 2 A could be used. In either case it is possible to match the current capacity of the SCR to the required load values.

Figure 5-12(b) shows another application in which SCR devices of low-current rating may be used to control large currents in a load. When the rectifiers are not conducting, the primary of transformer T_1 appears as a high reactance in series with the load. When the rectifiers conduct the secondary of the transformer is effectively short circuited, the impedance of the primary winding is very low, and load current is at a maximum. If the transformer has a 10-to-1 turns ratio, only 1/10 of the load current is handled by the rectifiers. The average value of load current may be controlled by varying the conduction angle of the controlled rectifiers.

5-12 INVERTERS

The function of an inverter circuit is to produce from a d-c source, an a-c voltage that has either a fixed or variable frequency. Series-parallel banks of controlled rectifiers have been used in inverter circuits to control a-c motors delivering up to 6000 hp over wide speed-ranges. In low-power inverter-circuits, controlled rectifiers have been operated at 10,000 Hz and 10,000 v. Typical series and parallel inverter circuits appear in Fig. 5-13.

In the parallel inverter of Fig. 5-13(a), the current initially is zero, and the capacitor is uncharged. When SCR 1 conducts, current passes through inductor L and through half of the transformer winding from terminal 2 to terminal 3. Since the primary is center-tapped, an induced voltage appears across terminal 1 and 3. This voltage charges the capacitor C. An induced voltage also appears in the secondary, and produces current through load R_L.

While SCR 1 is conducting, a positive charge builds up on terminal B of the capacitor. When SCR 2 is fired, conduction occurs immediately because of the positive voltage applied by the capacitor on the rectifier anode. The negative charge on point A of the capacitor cuts off SCR 1, and the process reverses, with terminal A of the capacitor now charging to a positive voltage. When SCR 1 is again fired, the cycle repeats. An alternating voltage thus appears at the secondary terminals of the transformer.

In the series inverter circuit of Fig. 5-13(b) the induced voltage in the

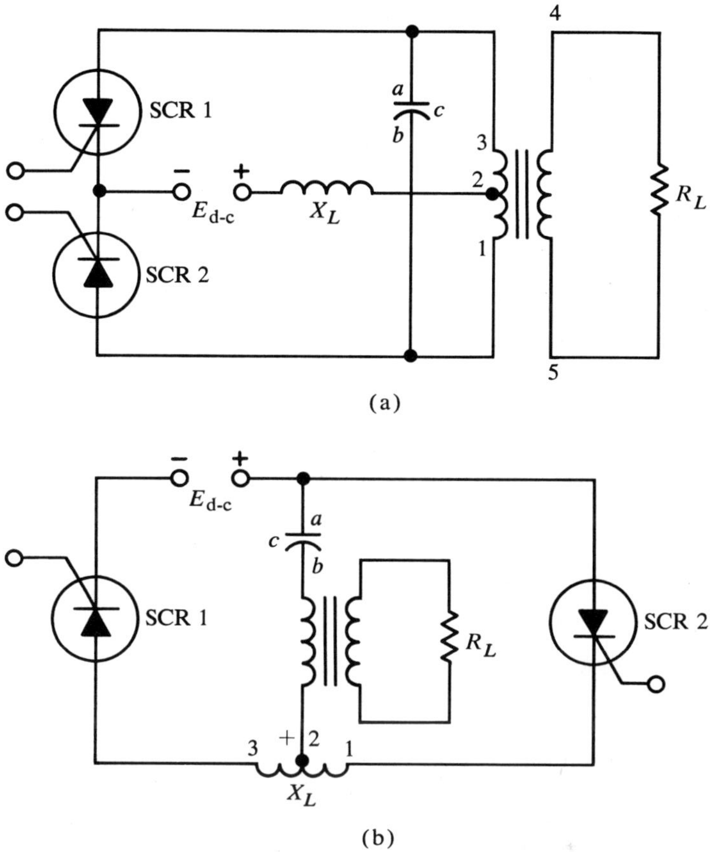

FIG. 5-13. (a) Parallel-inverter circuit; (b) Series-inverter circuit.

center-tapped inductor is used to cut off the controlled rectifiers. Initially the current in this circuit is assumed to be zero, and the capacitor is discharged. When SCR 1 is triggered, current passes through the primary of the transformer and also from terminal 2 to terminal 3 of the center-tapped inductor. The induced voltage in the inductor will cause point 2 to become positive with respect to point 3. After capacitor C has charged to the battery voltage, SCR 2 is triggered. The capacitor then discharges through SCR 2. The current through X_L induces a voltage which causes point 3 to become negative; SCR 1 is cut off. The inductive reactance X_L prevents more than one SCR from conducting at the same time because the induced voltage will always reverse-bias one or the other SCR. If both units were to conduct simultaneously, the power supply would be short-circuited.

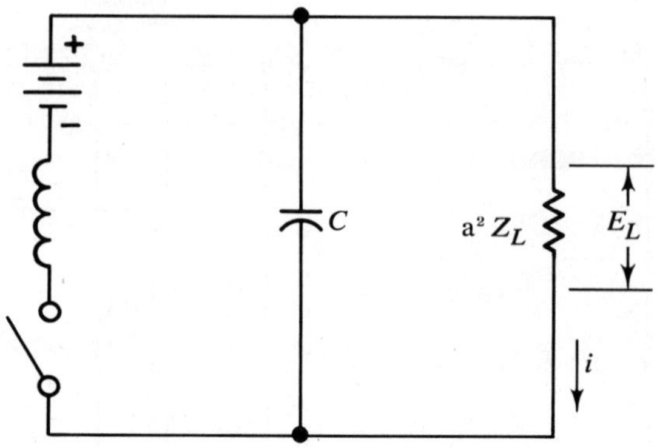

(a)

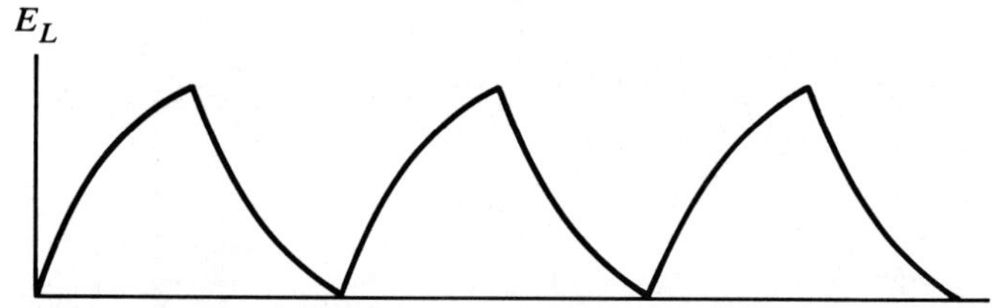

(b)

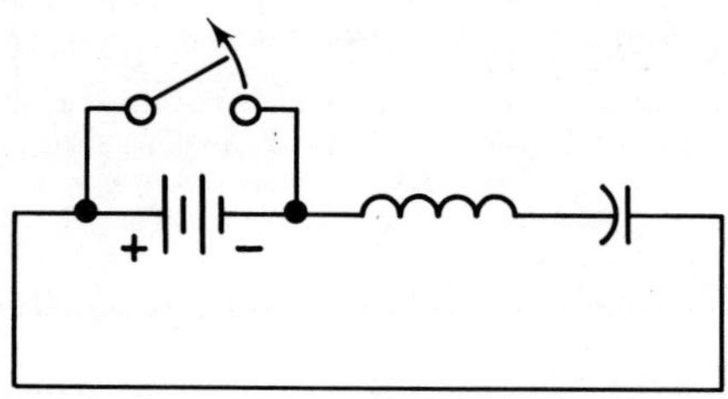

(c)

FIG. 5-14. (a) Equivalent circuit for parallel inverter; (b) Waveform of inverter output voltage; (c) Simplified inverter circuit.

5-13 CHARACTERISTICS OF SINGLE-PHASE INVERTERS

If the excitation current of the transformer is neglected, the parallel inverter circuit of Fig. 5-13 may be redrawn, as shown in Fig. 5-14. Ignoring the excitation current simplifies analysis of the inverter circuit. The load impedance appears to be directly across the capacitor C and also modified by the square of the transformer-turns ratio a. When the load is purely resistive and the value of R_L is low, the ratio of L/R determines the current waveform in the load. The voltage e_L across the load may then be expressed as

$$e_L \cong E(1 - E^{-Rt/L})$$

The waveshape appears as a series of transients, as shown in Fig. 5-14(b).

As R_L increases to infinity, the circuit is simplified, as shown in Fig. 5-14(c). The Kirchhoff loop equation for this circuit is

$$E = L\frac{di}{dt} + \frac{1}{C}\int i\,dt$$

Solving for the voltage across the capacitor, e_c

$$e_c = K_1 \cos(\omega t + \phi)$$

where

$$\omega t = \frac{1}{\sqrt{LC}}$$

It is apparent that the form of the load voltage is sinusoidal. As the value of the load resistor increases, the output waveform approaches the shape of a sine wave whose frequency is determined by L and C of the circuit. If the load were reactive instead of resistive, the waveform would again approach the form of a sine wave, and the frequency would be affected by the reactance in combination with the normal circuit reactances.

Problems

1. Plot a curve of the average d-c voltage across R_L of Fig. 5-5 vs the firing angle ϕ. Assume the firing angle to vary from zero to 180°. $E_{\text{in}} = 100$ v.
2. Plot a curve of the power in the load resistor R_L of Fig. 5-5 as the firing angle varies from zero to 180°. $E_{\text{in}} = 100$ v; $R_L = 20\ \Omega$.
3. $E_{\text{in}} = 100$ v and $R_L = 50$ ohms for the circuit of Fig. 5-5. What is the conduction angle when the power in the load is 30 W?
4. What is the maximum power that can be delivered to the load of Fig. 5-5 if $E_{\text{in}} = 500$ v and $R_L = 200\ \Omega$?
5. How much power will be delivered to the load of Problem 4 when the conduction angle is 30°?
6. For the half-wave SCR rectifier shown in Example 1, $R_L = 1000\ \Omega$, $E_{\text{in}} = 500$ v at 60 Hz, $R = 4000\ \Omega$, $C = 0.8\ \mu$F. Assume the SCR fires when V_{BD} becomes positive. Find the firing angle ϕ, the conduction angle β, and the average d-c load current.

7. Plot the firing angle vs the value of C in the circuit of Example 5-1 for variations in C from 0.1 to 20 μF.
8. For the full-wave SCR circuit of Fig. 5-6, $E_{in} = 220$ v and $R_L = 20\ \Omega$. What is the maximum power that can be delivered to this load? What firing angle produces that maximum power?
9. What firing angle will produce a load power of 500 W for the circuit of Problem 8?
10. Calculate the average voltage across the load of the SCR circuit of Fig. 5-7 when $E_m = 200$ v, $R = 1000\ \Omega$, $C = 30\ \mu$F, and the firing angle is 130°.
11. What would the ripple factor be for the circuit of Problem 10?
12. For the circuit shown in Fig. 5-7 $E_m = 200$ v, $R_L = 500\ \Omega$, and $C = 40\ \mu$F. Find the average SCR current during the conduction period for a firing angle of 60°.
13. The circuit of Fig. 5-10 has an input voltage of 220 v and a 1500-ohm load. Find the rms voltage across the load at $\phi =$ (a) zero°, (b) 60°, and (c) 120°.
14. What must the vA rating and turns ratio of a transformer be if it is to be used to match two 10-A SCR devices to a 220-v, 10,000-W load?
15. What will be the rms value of the voltage across the load in the circuit of Fig. 5-10(a) when $E_{in} = 220$ v, $R = 500\ \Omega$, and $\phi =$ (a) zero°, (b) 60°, (c) 120°, and (d) 180°?

Questions

1. How is the average current in a thyratron circuit controlled?
2. What limits the frequency of operation of a thyratron tube?
3. State the advantages of the SCR over the thyratron.
4. Why can an SCR continue to conduct after its gate current is reduced to zero?
5. How is the conduction of current through an SCR reduced to zero?
6. How will the magnitude of the supply voltage across an SCR affect the current required for firing?
7. How must an SCR-firing circuit differ from a thyratron-firing circuit?
8. If the resistor R in Fig 5-4(a) is decreased, how will the average current in R_L be affected? Why?
9. What will be the average load current in the SCR circuit of Fig. 5-5 for firing angles of (a) zero°, (b) 90°, and (c) 180°?
10. Define the conduction angle and firing angle of an SCR rectifier circuit.
11. What will be the average load current in the SCR circuit of Fig. 5-6 if the firing angle is (a) zero°, (b) 90°, and (c) 180°?

12. What factors determine the angle at which an SCR will conduct in an SCR circuit incorporating a resistance-capacitance load?
13. Explain why the conduction angle can affect the peak SCR current in an SCR circuit with *RC* load.
14. If the SCR of Fig. 5-7 is fired before the input voltage and capacitor voltage are equal, what will determine the peak current in the SCR?
15. How would an *RC* load affect the magnitude of the current required to fire an SCR? Why?
16. Why is Eq. 5-11 not valid if the firing angle of the SCR is greater than 90°?
17. Modify Eq. 5-11 so that it will be valid for known firing angles greater than 90°.
18. Describe mode 1 and mode 2 operation for a full-wave SCR bridge circuit with an inductive load.
19. What are the advantages of controlling power to a load by means of the circuit of Fig. 5-10 instead of by a simple series rheostat?
20. Explain why the load current in the circuit of Fig. 5-11 controls the firing point of each controlled rectifier when the circuit is operating in mode 1.
21. What determines the frequency of the a-c voltage produced by the inverter circuit of Fig. 5-13(a) and 5-13(b)?

Regulators

6

A regulator is a control device that automatically holds some physical quantity, such as pressure, speed, acceleration, voltage, or other variable, to a desired value. In general, a regulated control system consists of (a) a sensing device that detects any change in the variable being controlled and (b) a control circuit that is activated by a signal from the sensing device. The sensing device continuously compares the output quantity to a reference, and any error (difference between the reference and the output) is used to activate the control circuit. The control circuit then acts in a way to reduce the difference between the reference and the output.

In Fig. 6-1, a block diagram of a typical d-c-power supply regulator is presented. The output voltage is continuously compared to a reference, and any difference or error between the two quantities is fed back to the control element. Circuits of this type are commonly known as *feed back*-control circuits. Feedback control systems always operate by balancing a reference level and an output quantity. An error signal, obtained by comparing the reference with the output is used to drive a system that corrects the output level and reduces the error to zero.

A simple regulating device is the household thermostat. A reference level is set by turning a thermostat to some desired temperature. When the room temperature is far enough below this level, an error signal is sent to the furnace control, and the heating system is activated. When the room temperature is far enough above the reference level, an error signal of the opposite polarity is sent to the furnace or boiler control, and the furnace is shut down by the control system. In continuously trying to maintain a zero error, the thermostat balances the room temperature with the reference level.

Every regulation system must have an error-sensing device that compares the output quantity to a reference. It must also have some means of using this error signal to adjust the output so that the error is reduced and approaches zero.

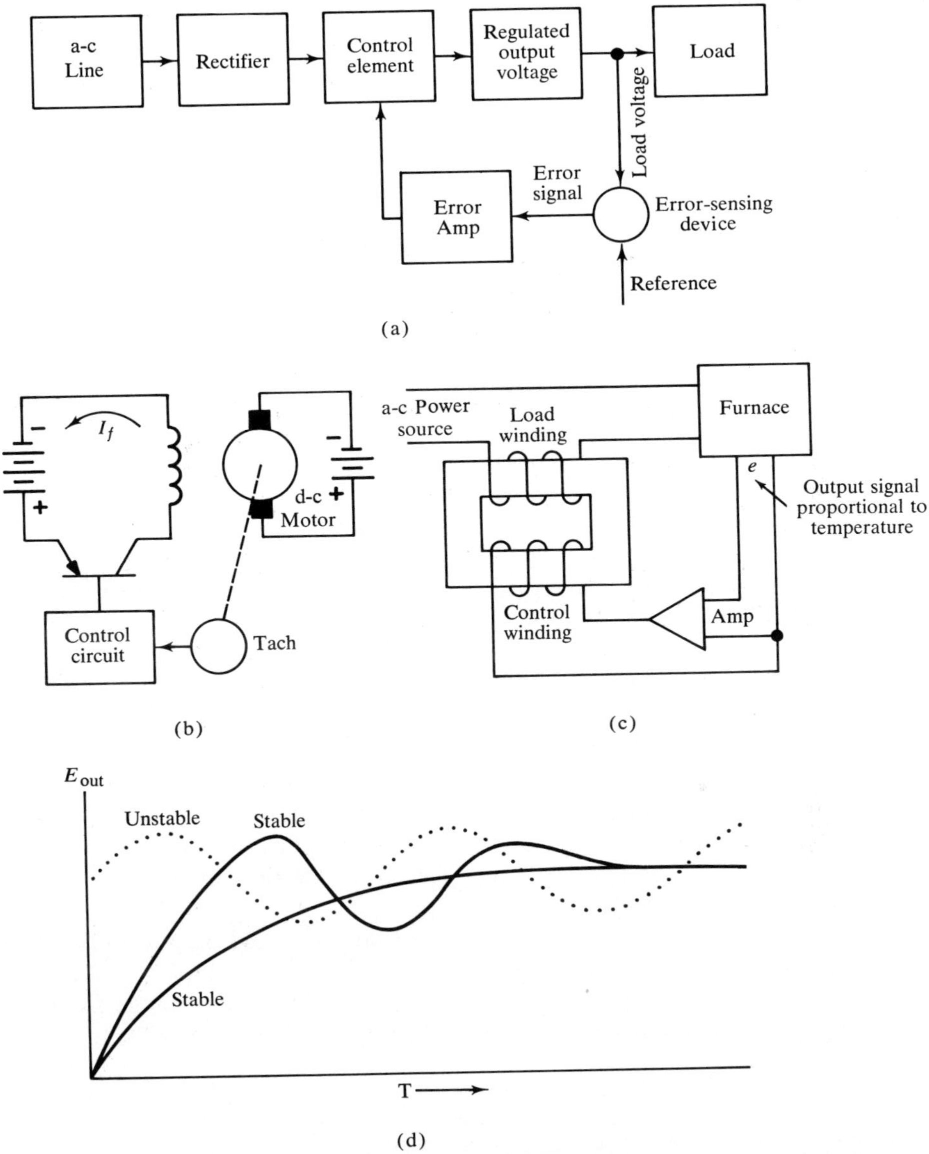

FIG. 6-1. (a) Block diagram of regulated power supply; (b) Schematic for speed-regulation system; (c) Saturable reactor used in temperature control system; (d) Output voltage vs time characteristics for typical regulation circuits.

In Fig. 6-1(b) the schematic drawing for a speed-regulation system is shown. In the system illustrated a transistor is used to control the current passing into the field of a d-c motor. The output of the tachometer is compared to a reference in the control circuit, and the error between the tachometer output and the reference is used to drive the control circuit and adjust the motor-field current.

Figure 6-1(c) shows the way in which a saturable reactor could be used to regulate the temperature of a furnace. The thermocouple voltage is fed into a bridge circuit and compared with a zener-diode reference voltage. Any error voltage is amplified and used to drive the control winding of the saturable reactor.

One important parameter of a regulation system that uses feedback is its *response time*, the time required to reduce the error to zero after the output or reference is changed. A second is its percentage of regulation, or the difference between the output and reference that always exists in the steady state. A third important characteristic, the stability of the system, is a function of the number of times the output level will oscillate above and below the reference level, after a change in output has occurred, before settling to its final value. In an unstable system the output may continually oscillate, never settling to a final value. A very stable system would be one in which the variable quantity reaches its final value without overshooting. Figure 6-1(d) shows output vs time curves for various systems.

The heart of any voltage-regulating system is the voltage-sensing circuit which constantly monitors the output voltage and detects any variation from the desired value. This circuit must be very sensitive and detect minute changes in voltage if close regulation is to be obtained.

The VR tube, previously discussed, may be used as an error-detecting device at the output of a power supply. The constant voltage characteristic of this tube makes it useful as the voltage reference in an error-detecting circuit. A circuit, using the VR tube for error-detection, appears as Fig. 6-2 and is analyzed in subsequent paragraphs.

The detection of very small changes in power-supply output voltage can be accomplished by use of bridge circuits. A circuit of this type, which is used in many power supplies, is shown in Fig. 6-4. The error voltage from the detector circuit is zero as long as the voltages in the bridge have the following relationship:

$$\frac{V_{\text{zener}}}{V_3} = \frac{V_1}{V_2}$$

If E_{in} from the power supply varies, the error voltage will become plus or minus, depending upon whether E_{in} increases or decreases. A feedback cir-

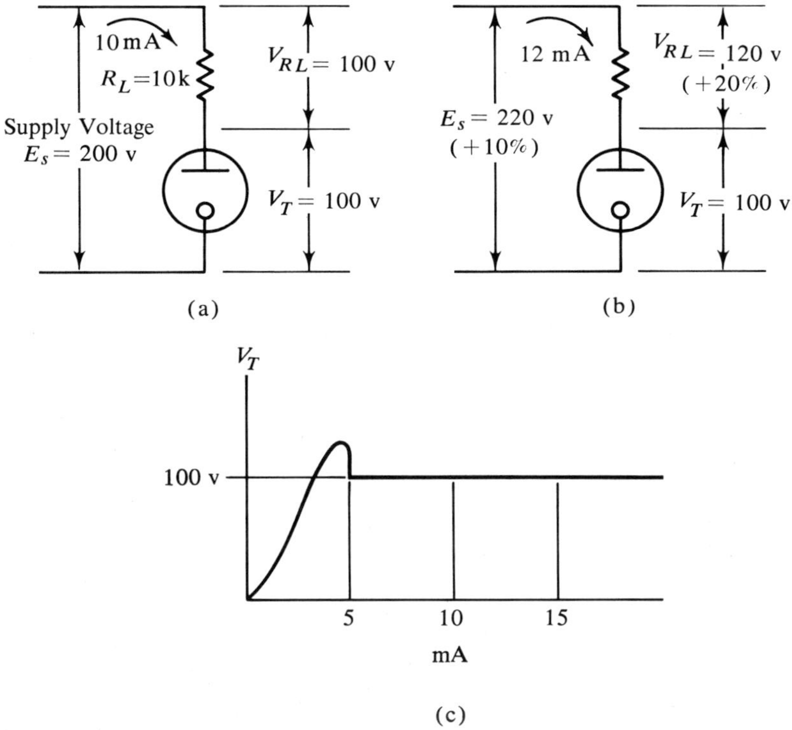

FIG. 6-2. (a) Error-detection circuit using VR tube, nominal input voltage of 200 v; (b) Input increased by 10 percent; (c) Typical characteristics of VR tube.

cuit will continuously adjust E_{in} to keep the error voltage at zero. Since this circuit is extremely sensitive, minute changes in output voltage will result in very large percentage changes in error voltage.

Voltage-sensing devices may be mechanical, thermal, or magnetic, depending upon the design of the system, but for every type of device the sensitivity, or ability to detect small voltage changes, is of prime importance.

6-1 VOLTAGE-REFERENCE DEVICES

The voltage-regulation circuit shown in Fig. 6-1 requires a stable reference voltage for proper operation. After the output voltage of the power supply is compared to the reference voltage, any difference in magnitude is amplified and fed back to the control element. When regulation within a fraction of one percent is required, the reference voltage must be very stable. An error-sensing circuit must also respond to minute changes in the power supply output.

The voltage-regulator gas tube described in Chapter 4 is often used as a voltage-reference device. An error-detecting (sensing) circuit using a VR tube is illustrated in Fig. 6-2. The characteristic curve in Fig. 6-2(c) shows that the voltage drop across the tube is a constant 100 v when the tube current varies from 5 to 15 mA. This constant voltage characteristic of the VR tube makes it suitable for use in a power-supply voltage-sensing circuit. The output voltage of the power supply is used as the input voltage of the sensing circuit. Since the drop across the VR tube is constant, any change in the input voltage appears across the load resistor R_L. The voltage across R_L is then fed back into the control circuit and permits E_{out} to remain constant.

The circuit of Fig. 6-2 also allows amplification of the percentage of change in the input voltage. The entire change in the input voltage appears across R_L. If the nominal supply voltage of 200 v rises by 10 percent or 20 v, the 20-v change will appear across R_L. These 20 v represent a 20 percent change in the nominal value of voltage across R_L. The change in input voltage has, therefore, been amplified by a factor of two.

Since the entire change in power-supply voltage appears across the load resistor,

$$\text{Percent change of } E_s = \frac{\Delta E_s}{E_s} \times 100$$

and

$$\text{Percent change of } V_{R_L} = \frac{\Delta E_s}{V_{R_L}} \times 100$$

Also,

$$(100)\Delta E_s = E_s(\text{percent change of } E_s)$$

and

$$(100)\Delta E_s = V_{R_L}(\text{percent change of } V_{R_L})$$

Setting these two equations equal to each other,

$$\frac{E_s}{V_{R_L}} = \frac{\text{percent change of } V_{R_L}}{\text{percent change of } E_s} \tag{6-1}$$

Now, let this relationship be used in designing a voltage-sensing circuit for a power supply.

Example 6-1 Design an error-sensing circuit using a VR tube, with the characteristics shown in Fig. 6-2, that can be used to amplify the percent of change in a 250-v power supply. Calculate the percent of change in output voltage for this circuit resulting from one percent change in the power supply voltage. Since the VR tube shown will maintain 100 v from anode to cathode, two tubes in series are used to obtain maximum error-amplification. Using

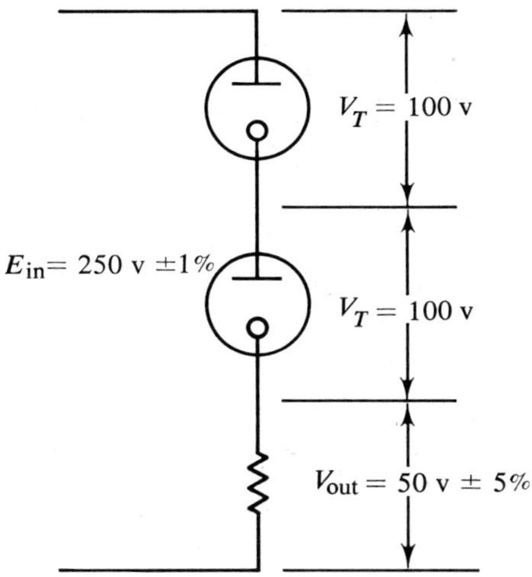

FIG. E6-1. Circuit for Example 6-1.

Eq. 6-1 the percent change in sensing-circuit output can be calculated for one percent variation in supply voltage.

$$\frac{250}{50} = \frac{\text{Percent change of } V_{out}}{1 \text{ percent}}$$

$$\text{Percent change } V_{out} = 5$$

This circuit provides an error-amplification factor of 5.

The zener diode can also be used as a power-supply voltage reference. A voltage reference circuit, using a zener diode, and a typical zener characteristic curve are shown in Fig. 6-3. The zener maintains a constant voltage drop from anode to cathode in the breakdown region and can, therefore, be used as a voltage reference. More rugged than the VR tube, the zener is produced in a wide range of current and voltage ratings. However, the breakdown voltage does vary with temperature, and compensation techniques may be required wherever close regulation is desired.

6-2 VOLTAGE-REFERENCE AND ERROR-DETECTION BRIDGE CIRCUITS

A high degree of accuracy in voltage regulation may be obtained by using a bridge circuit for error-detection. The advantage of this type of circuit is

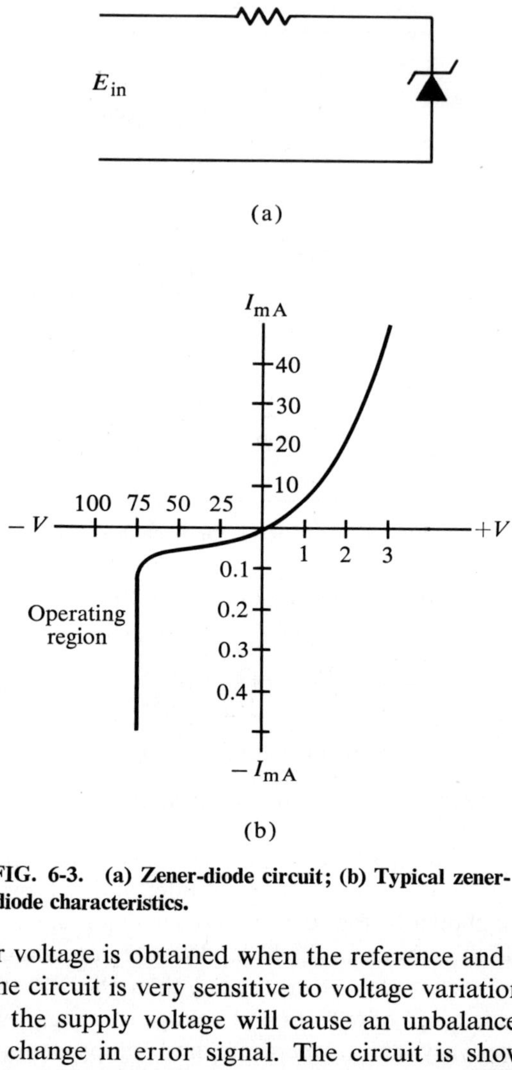

FIG. 6-3. (a) Zener-diode circuit; (b) Typical zener-diode characteristics.

that a zero-error voltage is obtained when the reference and supply voltages are balanced. The circuit is very sensitive to voltage variations, since even a slight change in the supply voltage will cause an unbalance in the bridge, causing a large change in error signal. The circuit is shown in Fig. 6-4. This circuit has the added advantage of producing a positive or negative error voltage when the supply voltage rises or falls. The expression for the error voltage can be found by writing the loop equation:

$$E_{in} = V_Z + \text{Error} + R_2 I_1$$

If the current in the center leg of the bridge is zero,

$$I_1 = \frac{E_{in}}{R_1 + R_2}$$

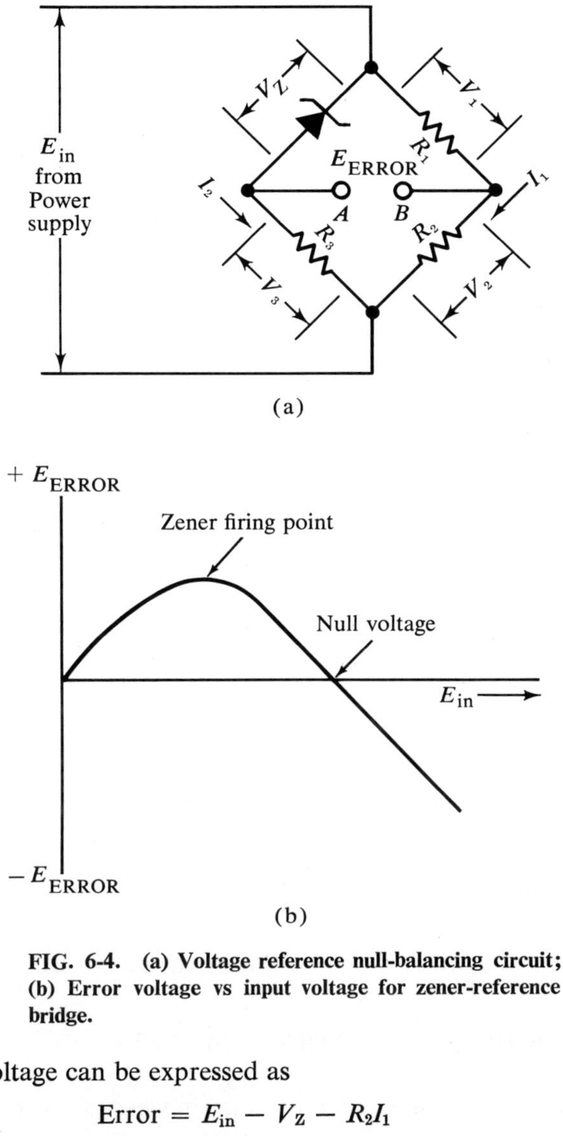

FIG. 6-4. (a) Voltage reference null-balancing circuit; (b) Error voltage vs input voltage for zener-reference bridge.

The error voltage can be expressed as

$$\text{Error} = E_{in} - V_Z - R_2 I_1$$

from which

$$\text{Error} = E_{in} - V_Z - \frac{E_{in}(R_2)}{R_1 + R_2}$$

and

$$\text{Error} = E_{in}\left(1 - \frac{R_2}{R_1 + R_2}\right) - V_Z \qquad \textbf{(6-2)}$$

From Eq. 6-2 it can be seen that V_Z is the reference voltage and that the error voltage may be plus, minus, or zero, depending upon the value of E_{in}. A plot of the error voltage vs E_{in} is shown in Fig. 6-4(b).

Example 6-2 Using the zener diode whose characteristics are shown in Fig. 6-3, design a voltage-sensing bridge that can be used to detect the error in the output of a 250-v power supply. Calculate the magnitude and the polarity of the error voltage for a one-percent change in input.

Referring to Fig. 6-4, it is noted that for zero-error voltage the voltage drop across R_2 must equal the drop across R_3. The voltage across the zener is 75 v in the breakdown region, and 2 mA is selected as the nominal zener current. Under these conditions

$$I_2R_3 = I_1R_2 = 175 \text{ v}$$

and

$$V_Z = I_1R_1 = 75 \text{ v}$$

Now, if

$$I_1 = I_2$$

then

$$R_2 = R_3 = 175/(2 \times 10^{-3}) = 87.5 \text{ k}$$

and

$$R_1 = 75 \text{ v}/(2 \times 10^{-3}) = 37.5 \text{ k}$$

For a one-percent change in supply voltage $E_{in} = 252.5$ v. The error voltage can now be calculated from Eq. 6-2.

$$\text{Error} = 252.5\left(1 - \frac{87.5}{125}\right) - 75$$

$$\text{Error} = +0.75 \text{ v}$$

The error is plus, or point A is plus with respect to point B.

6-3 THE ZENER DIODE AS A VOLTAGE-REGULATOR

A simple regulating circuit may be constructed in which a zener diode is used as a basic voltage regulator. Such a circuit can give satisfactory performance in circuits providing for control of small amounts of power and not requiring accurate regulation. One source of error in a zener regulator, such as that shown in Fig. 6-5, is the variation of breakdown voltage with temperature. A 5-v zener may have a variation in breakdown voltage of 0.01 percent/°C. As the breakdown voltage rises, this variation increases and a 30-v zener may

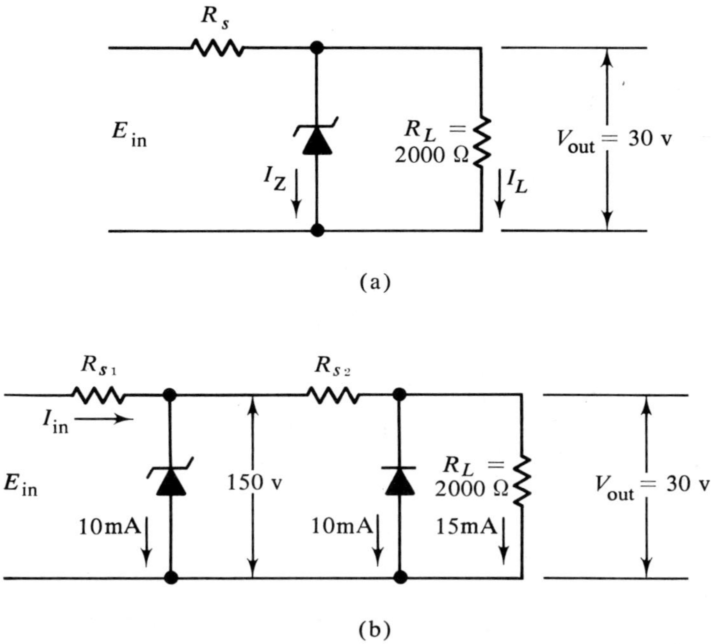

FIG. 6-5. **(a) Zener-diode-regulation circuit; (b) Cascade regulation circuit.**

have a voltage coefficient of 0.06 percent/°C. If close regulation is required, six 5-v zeners in series will have better temperature characteristics than one 30-v zener.

The zener-regulating circuit of Fig. 6-5 is basically the same as the VR circuit discussed in Section 4-2. Example 6-3 is presented here as a review of the method used to analyze the zener-regulating circuit.

Example 6-3 Over what range of input voltages will the zener-regulating circuit shown in Fig. 6-5 maintain 30 v across the 2000-Ω resistor. Maximum zener current is not to exceed 25 mA, $R_s = 200\ \Omega$.

Load current I_L will be 15 mA, and if the zener current is zero (just before breakdown),

$$E_{in} = R_s I_L + V_Z$$
$$E_{in} = 200(0.015) + 30$$
$$E_{in} = 33 \text{ v}$$

When the zener current is at the maximum value of 25 mA,

$$E_{in} = 200(0.025 + 0.015) + 30$$
$$= 38 \text{ v}$$

The output voltage will be maintained at 30 v when the input varies from 33 to 38 v.

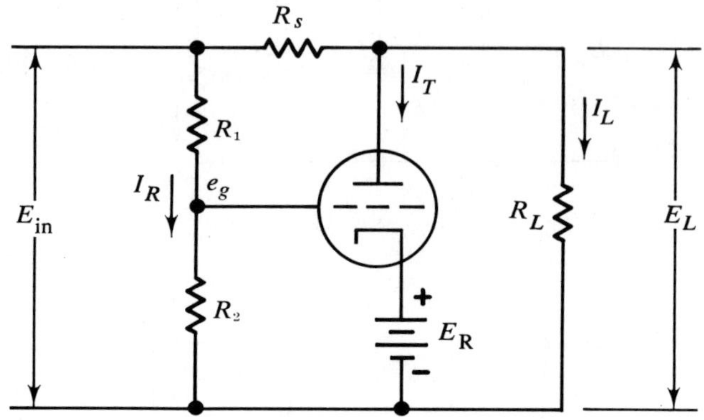

(a)

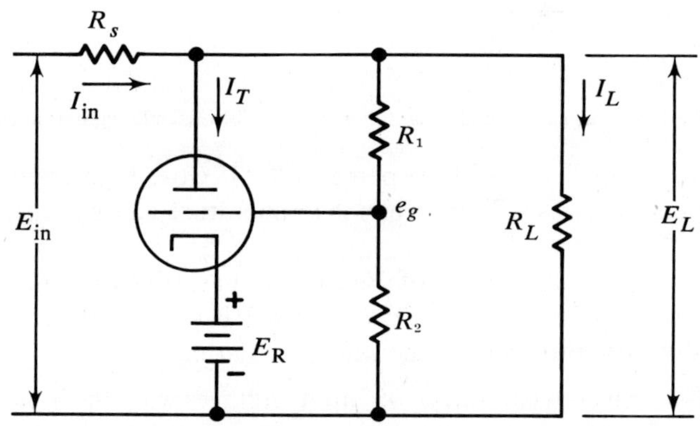

(b)

FIG. 6-6. (a) Line-regulation circuit without feedback from load voltage (this circuit corrects for changes of input voltage only); (b) Line-regulation circuit with feedback from load voltage (this circuit corrects for changes of input voltage and load voltage).

When the regulated voltage is much lower than the supply voltage, the two-stage zener-regulator circuit of Fig. 6-5(b) gives a reduction of the power loss in the series-dropping resistor. If E_{in} equals 200 v and the nominal zener current is 20 mA, for the regulation circuit shown in Fig. 6-5(a), the voltage drop across the series resistor must be 170 v and the power loss may be calculated as

$$P_s = 170 \text{ v} \times (0.015 + 0.020) = 5.95 \text{ W}$$

If the circuit of Fig. 6-5(b) is used and a 150-v zener placed ahead of the 30-v zener, the values of R_{s1} and R_{s2} can be chosen so that each zener conducts 10 mA; the power loss is calculated as follows

$$R_{s1} = 50 \text{ v}/35 \text{ mA} = 1430\,\Omega$$

and

$$R_{s2} = 120 \text{ v}/25 \text{ mA} = 4800\,\Omega$$

$$(\text{power lost})\ R_{s1} = 50(0.035) = 1.75 \text{ W}$$

$$(\text{power lost})\ R_{s2} = 120(0.025) = 3 \text{ W}$$

$$(P_s \text{ total}) = 4.75 \text{ W}$$

which is more than a 20-percent reduction in loss. Further cascading of zener diodes can bring additional reduction in power loss.

6-4 FEEDBACK-VOLTAGE REGULATORS

In the closed loop or feedback regulator, for which the block diagram was shown in Fig. 6-1(a), the output voltage is constantly monitored, and any output voltage change is fed back to the control circuit for correction. In Fig. 6-6, both open- and closed-loop regulation circuits are presented.

In the circuit of Fig. 6-6(b), the grid voltage is directly proportional to the load voltage; therefore, the load voltage determines the tube current and the voltage drop in R_s. In this circuit, which provides automatic voltage regulation at the load, the tube is able to sense changes of load voltage due either to input voltage change or load current change. If it is assumed that R_1 and R_2 are so large that the current drawn by the network is negligible, the circuit of Fig. 6-6(b) can be analyzed as follows:

$$e_g = E_L \frac{R_2}{R_1 + R_2}$$

if

$$\frac{R_2}{R_1 + R_2} = \beta$$

then

$$e_g = \beta E_L$$
$$I_{\text{in}} = I_T + I_L \tag{6-3}$$

which may be rewritten as

$$\frac{E_{\text{in}} - E_L}{R_s} = \frac{E_L + \mu e_g}{R_P} + \frac{E_L}{R_L} \tag{6-4}$$

Now, combining Eqs. 6-3 and 6-4,

$$\frac{E_{\text{in}} - E_L}{R_s} = \frac{E_L + \mu\beta E_L}{R_P} + \frac{E_L}{R_L} \tag{6-5}$$

Rearranging Eq. 6-5,

$$\frac{E_{\text{in}}}{R_s} = E_L\left(\frac{1}{R_s} + \frac{1}{R_P} + \frac{1}{R_L} + \frac{\mu\beta}{R_P}\right) \tag{6-6}$$

from which

$$E_{\text{in}} = E_L\left(1 + \frac{R_s}{R_P} + \frac{R_s}{R_L} + \frac{\mu\beta R_s}{R_P}\right) \tag{6-7}$$

Rearranging Eq. 6-7 and substituting $g_m = \mu/R_P$

$$E_L = \frac{E_{\text{in}}}{1 + \dfrac{R_s}{R_P} + \dfrac{R_s}{R_L} + g_m\beta R_s} \tag{6-8}$$

Since the load voltage changes with respect to the input voltage, E_L of Eq. 6-8 is differentiated with respect to E_{in}.

$$\frac{\partial E_L}{\partial E_{\text{in}}} = \frac{1}{1 + \dfrac{R_s}{R_L} + \dfrac{R_s}{R_P} + g_m\beta R_s} \tag{6-9}$$

To obtain the expression for load voltage changes with respect to changes in load resistance, rearrange Eq. 6-8 and differentiate E_L with respect to R_L.

$$\frac{\partial E_L}{\partial R_L} = \frac{E_L(R_s/R_L{}^2)}{1 + \dfrac{R_s}{R_L} + \dfrac{R_s}{R_P} + g_m\beta R_s} \tag{6-10}$$

Equation 6-9 gives the expression for the change in E_L with respect to E_{in}. If the regulating tube were disconnected, Eq. 6-9 would reduce to

$$\frac{\partial E_L}{\partial E_{\text{in}}} = \frac{1}{1 + \dfrac{R_s}{R_L}} \tag{6-11}$$

Comparing Eq. 6-11 with Eq. 6-9 reveals that the change in E_L with E_{in} is improved by increasing the denominator of Eq. 6-11 by $R_s/R_P + g_m\beta R_s$.

Equation 6-10 shows the variation of E_L with respect to changes in R_L. If the regulating circuit were disconnected, Eq. 6-10 would reduce to

$$\frac{\partial E_L}{\partial R_L} = \frac{E_L(R_s/R_L{}^2)}{1 + \dfrac{R_s}{R_L}} \tag{6-12}$$

A comparison of Eqs. 6-10 and 6-12 shows that the variation of E_L with respect to load changes is improved by increasing the denominator of Eq. 6-12 by the same factor of $R_s/R_P + g_m\beta R_s$. Therefore, the regulation circuit has decreased the amount of change in output voltage due to change either in input voltage or load.

6-5 THE TRANSISTORIZED SHUNT-REGULATOR CIRCUIT

The vacuum-tube circuit discussed in Section 6-4 can be transistorized, as shown in the circuit of Fig. 6-7(a). Analysis of this shunt-regulation circuit permits determination of the factors that govern variations in output voltage.

In Fig. 6-7(b) the transistor of Fig. 6-7(a) is replaced by an equivalent circuit containing the collector resistance R_c, the base resistance R_b, and the emitter resistance R_e. In shunt with the equivalent collector resistance is the current generator βI_b, β being the current gain of the transistor and I_b the base current. The performance of this circuit may be evaluated by finding the ratio of the change in load voltage to the change in input voltage, or $\Delta VL/\Delta E_{\text{in}}$. Analysis of this circuit will disregard the steady-state d-c voltages and deal only with the variable or a-c component of the voltages ΔE and ΔVL.

The change in input voltage may be written as

$$\Delta E_{\text{in}} = \Delta V_s + \Delta V_L \tag{6-13}$$

and the ratio of load voltage change to input voltage changes as

$$\frac{\Delta V_L}{\Delta E_{\text{in}}} = \frac{\Delta V_L}{\Delta V_L + \Delta V_s} \tag{6-14}$$

Equation 6-14 may be rearranged and written as

$$K_V = \frac{\Delta V_L}{\Delta E_{\text{in}}} = \frac{1}{1 + \dfrac{\Delta V_s}{\Delta V_L}} \tag{6-15}$$

The ratio of ΔV_L to ΔE_{in}, a measure of the regulation of the circuit, is called the voltage-regulation constant; it is represented by the symbol K_V. It is

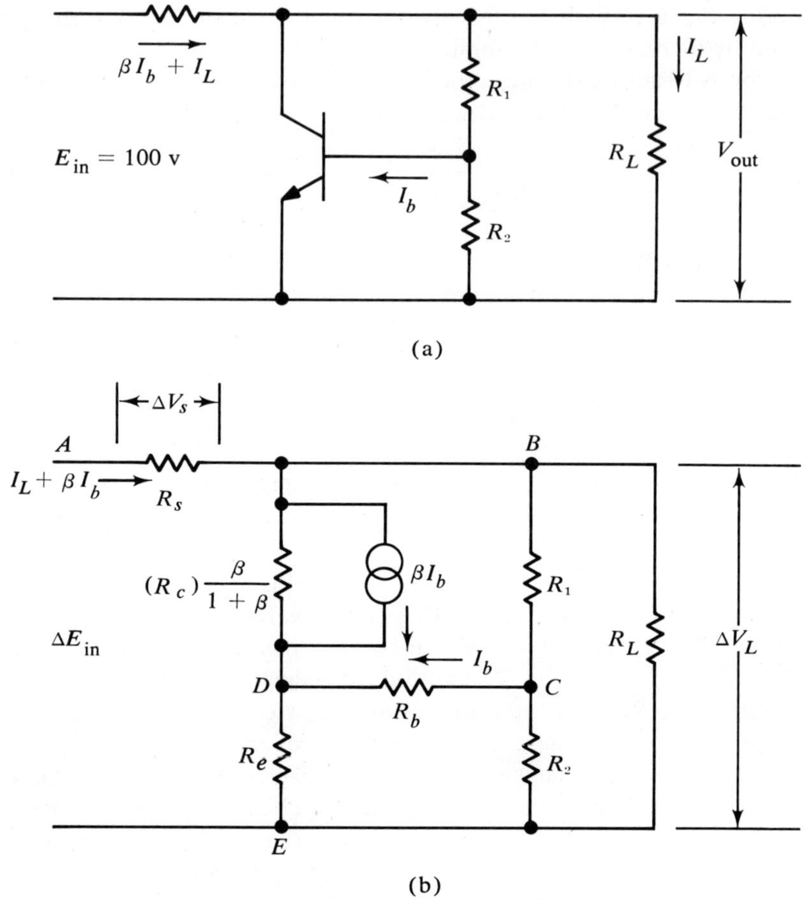

FIG. 6-7. (a) Transistorized shunt-regulation circuit; (b) Circuit equivalent to that in (a).

desirable to express the value of K_V in terms of the circuit parameters. This expression of K_V is found as follows:

From the equation for the voltage drops around loop B, C, D, E, the change in load voltage ΔV_L may be expressed as

$$\Delta V_L = R_1 \Delta I_b + R_b \Delta I_b + R_e(\beta \Delta I_b + \Delta I_b) \tag{6-16}$$

Because $I_b \ll 1$ and $R_b \ll R_1$, we may rewrite Eq. 6-16 as

$$\Delta V_L \cong R_1 \Delta I_b + R_e \beta \Delta I_b = \Delta I_b(R_1 + \beta R_e) \tag{6-17}$$

the voltage change across the series resistor ΔV_s is

$$\Delta V_s = R_s(\beta \Delta I_b + \Delta I_L) \tag{6-18}$$

Now writing ΔI_L in terms of ΔI_b:

$$\Delta I_L = \frac{\Delta V_L}{R_L}$$

and

$$\Delta I_b = \frac{\Delta V_L}{R_1 + R_b + R_e}$$

Combining these two equations, ΔI_L may be written as

$$\Delta I_L = \frac{\Delta I_b(R_1 + R_b + R_e)}{R_L}$$

Substituting this expression for ΔI_L into Eq. 6-18,

$$\Delta V_s = R_s \Delta I_b \left(\beta + \frac{R_1 + R_b + R_e}{R_L}\right)$$

Using this expression for ΔV_s and the expression for ΔV_L from Eq. 6-17, the voltage-regulation constant K_V is defined as follows:

$$K_V = \frac{1}{1 + \dfrac{\Delta V_s}{\Delta V_L}}$$

$$K_V = \frac{1}{1 + \dfrac{R_s\left(\beta + \dfrac{R_1 + R_b + R_e}{R_L}\right)}{R_1 + \beta R_e}} \tag{6-19}$$

which reduces to

$$K_V = \frac{1}{1 + \dfrac{\beta R_s R_L + R_1 + R_b + R_e}{R_L(\beta R_e + R_1)}} \tag{6-20}$$

For close regulation, the term

$$\frac{\beta R_s R_L + R_1 + R_b + R_e}{R_L(\beta R_e + R_1)}$$

should be much greater than 1. If the current gain β is high, this factor is greater than one, and good regulation is obtained.

Example 6-4 The following data are available for the circuit of Fig. 6-7(a). $R_1 = 1$ K, $R_s = 100\ \Omega$, $I_L = 0.2$ A, $I_b = 2$ mA, $\beta = 100$, $R_e = 3\ \Omega$, $R_L = 500\ \Omega$, $R_B = 10\ \Omega$.

Calculate the value of the voltage regulation constant K_V.

From Eq. 6-20

$$K_V = \frac{1}{1 + \dfrac{100(500)(100) + 1000 + 10 + 3}{500(1000 + 300)}}$$

$$K_V = \frac{1}{1 + 7.7}$$

$$K_V = 0.115$$

If $K_V = 0.115$, the output will change by 0.115 v for every 1-v change in the input.

6-6 THE SERIES-REGULATOR CIRCUIT

The shunt regulator has the disadvantage of high power-loss in the series voltage-dropping resistor. This high power-loss can be overcome by use of the series-regulator circuit shown in Fig. 6-8. Here, the collector-to-base resistance of transistor 2 is controlled by the divider network, made up of resistors R_1 and R_2. When the output voltage increases, the base current of transistor 1 also increases, causing the collector of transistor 1 to become more negative. The base current of transistor 2 will decrease. This results in an increase in the voltage drop across transistor 2. The load voltage will then be reduced until the circuit has returned to balance. In order to simplify the analysis of this circuit, part (a) of Fig. 6-8 is redrawn as part (b) of Fig. 6-8. Transistor 1 has been replaced by a generator with a current gain of -1. The proper polarity is thus provided to the base of transistor 2.

Looking only at the a-c component or change in voltages,

$$\Delta E_{\text{in}} = \Delta V_s + \Delta V_L$$

Writing the Kirchhoff equation around the loop $C, D, E, B, C,$

$$\Delta V_L = \beta \Delta I_b R_e + \Delta I_b R_b - I_1 R_2$$

Assume that $R_2 I_1 \ll \beta \Delta I_b R_e$.

Then

$$\Delta V_L = \Delta I_b(\beta R_E + R_b) \tag{6-21}$$

The change in voltage across the series transistor becomes

$$\Delta V_s = \beta \Delta I_b(R_c + R_E) \tag{6-22}$$

The voltage-regulation constant may now be computed.

$$K_V = \frac{\Delta V_L}{\Delta E_{\text{in}}} = \frac{\Delta V_L}{\Delta V_s + \Delta V_L}$$

or

$$K_V = \frac{1}{1 + \frac{\Delta V_s}{\Delta V_L}}$$

Substituting the expression for ΔV_L from Eq. 6-21 and the expression for ΔV_s from Eq. 6-22, K_V becomes

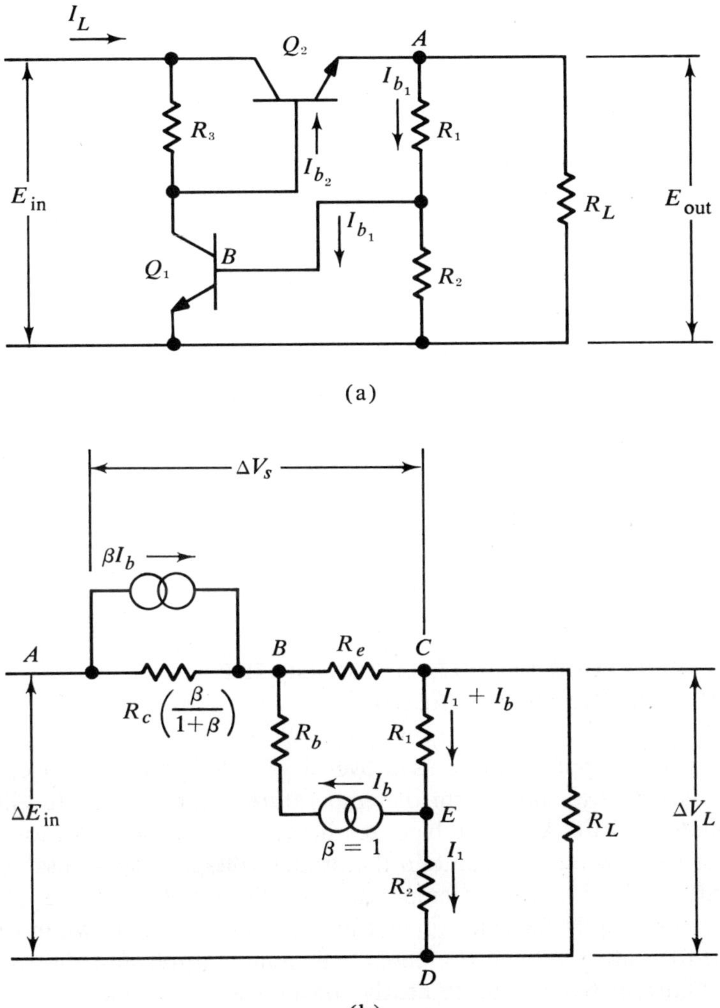

FIG. 6-8. (a) Transistorized-series-regulator; (b) Circuit equivalent to that in (a).

$$K_V = \frac{1}{1 + \dfrac{\beta \Delta I_b(R_c + R_e)}{\Delta I_b(\beta R_e + R_b)}}$$

which reduces to

$$K_V = \frac{1}{1 + \dfrac{\beta(R_c + R_e)}{\beta R_e + R_b}} \tag{6-23}$$

For good regulation the term

$$\frac{\beta(R_c + R_e)}{\beta R_e + R_b}$$

must be much greater than 1, which is true when current gain is high.

Example 6-5 For the circuit of Fig. 6-8(b), $\beta = 50$, $R_b = 10\ \Omega$, $R_c = 2000\ \Omega$, and $R_e = 2\ \Omega$. Find the voltage-regulation constant for this circuit.

From Eq. 6-23

$$K_V = \frac{1}{1 + \dfrac{50(2000 + 2)}{2 \times 50 + 10}}$$

$$K_V = \frac{1}{1 + \dfrac{100{,}000}{110}}$$

$$K_V \cong 0.001$$

6-7 TEMPERATURE COMPENSATION OF THE SHUNT-REGULATOR CIRCUIT

When a power supply is to be used over a wide temperature range, it may be necessary to use some form of temperature compensation to eliminate output-voltage drift. As the ambient temperature rises, the transistor will require less base current to maintain the proper voltage drop across the load. When an NTC or negative temperature coefficient-of-resistance device is placed in series with the zener current-limiting resistor, as shown in Fig. 6-9, the base-to-emitter voltage decreases with temperature, and the transistor collector current remains more nearly constant as the ambient temperature changes.

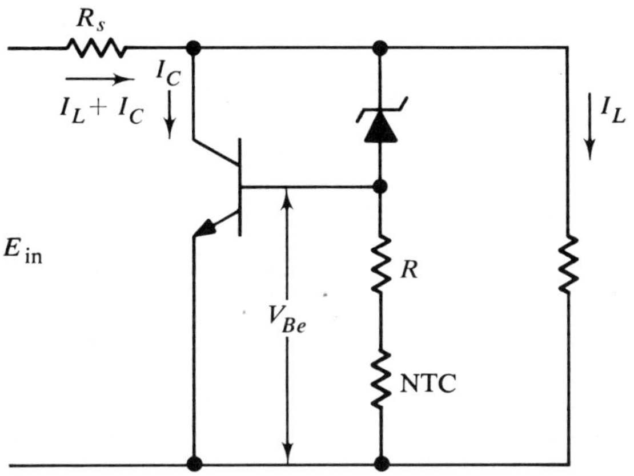

FIG. 6-9. Shunt-regulator circuit with temperature-compensating diode and NTC resistor.

6-8 THE REGULATED POWER-SUPPLY

A rectifier circuit with output-voltage regulation, such as that shown in Fig. 6-10, can be used to privide a stable d-c voltage source from an a-c line. A full-wave bridge circuit is used to rectify the a-c input, and capacitive filtering removes the a-c component from the output.

Notice that the output voltage is fed to one leg of a bridge circuit. A zener diode in the opposite leg of the bridge provides a reference voltage against which the output is compared. Any error voltage appearing at the bridge output is amplified and applied to the base of the control transistor where it acts to change the collector-to-emitter resistance of the transistor. This change, in turn, produces a change in the output voltage in such a manner as to achieve a balance between the output voltage and the zener reference voltage.

Current regulation may also be added to the d-c power supply, as shown in Fig. 6-11.

Voltage-regulation is achieved in the manner just described. However, current regulation is also obtained; the *IR* drop in a fixed resistor can be used as a reference voltage in the current-sensing bridge. The adjustable resistors in the voltage and current control legs of the bridges are used to set the crossover point. This is the point where voltage-regulation changes to current-regulation. When the *IR* drop in the series resistor reaches the preset value,

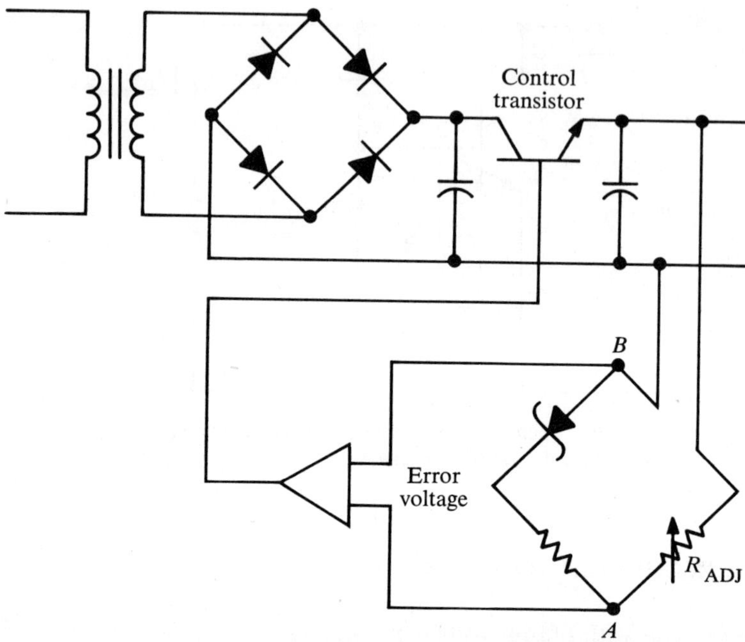

FIG. 6-10. Full-wave-regulated rectifier circuit with series transistor and zener-diode reference bridge.

the error voltage from the current bridge assumes control; the current then remains constant as the load changes.

The error from the zener bridge may also be used to trigger an SCR bridge as shown in Fig. 6-12.

6-9 REGULATED POWER-SUPPLY SPECIFICATIONS

The performance requirements of modern power supplies extend far beyond the requirement of simple voltage-regulation. Other factors such as output impedance, recovery time, stability are also important. The terms used to describe and specify power-supply performance deserve consideration.

Line-Regulation

This term is defined as the variation in output voltage that occurs when the input line voltage is varied from the minimum to maximum specified values, while the load current is held at 50 percent of rated output. The regulation is

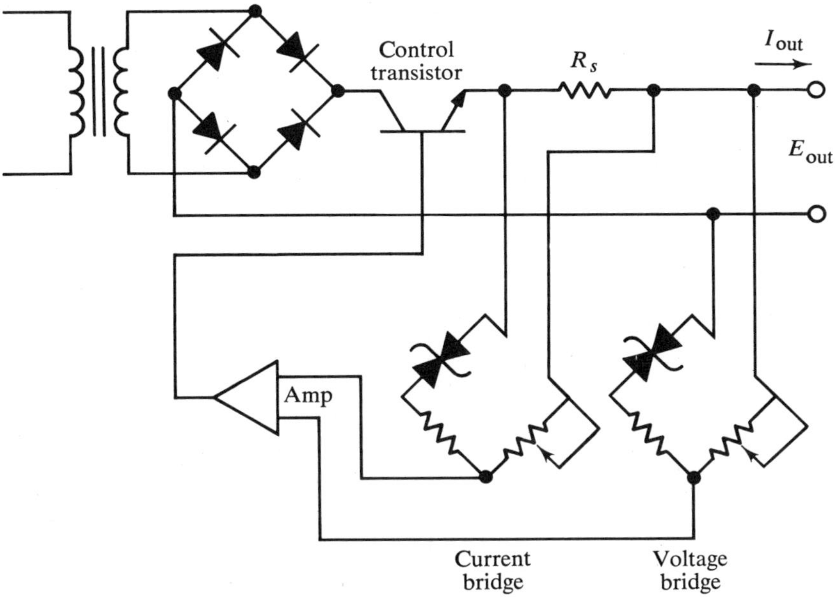

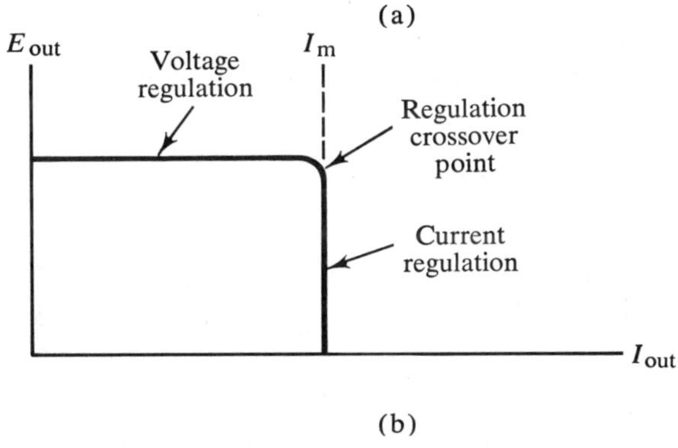

FIG. 6-11. (a) Full-wave bridge rectifier with voltage and current regulation; (b) Voltage vs current output characteristic.

specified as a percentage variation from a specified value. The regulation may also be expressed as the absolute value of the change in volts.

Example

The output voltage of power supply that delivers a nominal 200 v to a load varies from 197 to 203 as the input voltage is changed from 200 to 250. Express the regulation in terms of percent and absolute values.

$$200 \pm 1.5 \text{ percent} \quad \text{and} \quad 200 \text{ v} \pm 3 \text{ v}$$

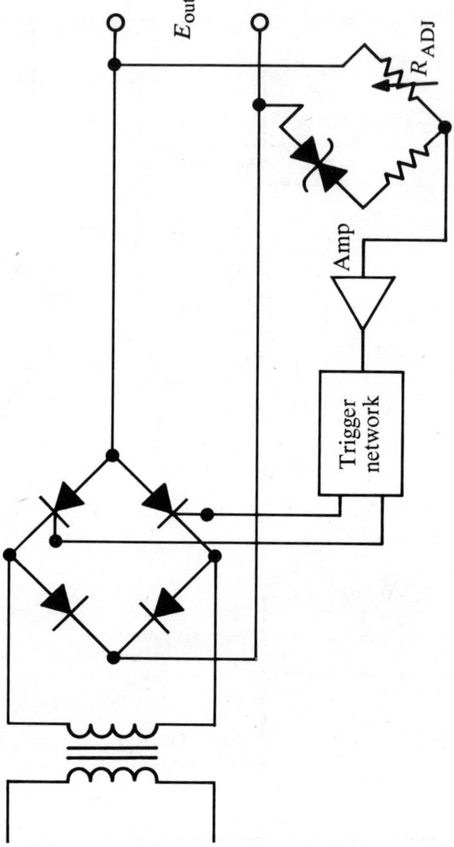

FIG. 6-12. Power supply using the SCR as the control element.

Load Regulation

Load-regulation is the variation in output voltage that occurs as the load current is varied from zero to rated value, while the input voltage is held at a nominal value. The regulation is expressed in terms of percent or as an absolute value in volts.

Total Regulation

Sometimes called total line-load-regulation, or total static regulation, total regulation is theoretically the worst possible combination of line and load variations. The term refers to the variation in output voltage caused by simultaneous variations in the line and load, resulting in output variations from each being added. If maximum input voltage produces maximum output voltage when the load is constant, and zero load current produces maximum output voltage when the input voltage is held constant, the total regulation is the output voltage at no-load with the input voltage at the maximum rated value.

Transient-Response or Dynamic-Load-Regulation

These terms describe the variation in output voltage that occurs when the load is abruptly changed by 25 percent from a base level of 50 percent of rated full load. The variation in output voltage disappears after a short time, and the output returns to its steady-state value. From Fig. 6-13 the transient-response is determined by the maximum overshoot value and the recovery time. The recovery time is the interval between the instant when the load is changed to the time when the output voltage returns and remains within the specified limits of static regulation. Maximum overshoot is the change in maximum output voltage from the average output value. The average value is calculated from the voltage output before and after the load change. Maximum overshoot may be expressed in percent or in absolute values.

Dynamic-Line-Regulation

With the load held constant at 50 percent of rated value, the input voltage is suddenly changed from the nominal value to either the maximum or minimum input value. The conditions for regulation are the same as those shown in Fig. 6-13 for dynamic-load-regulation.

Ripple (Periodic and Random Deviation)

Ripple includes only that portion of the output voltage that is a harmonic of the fundamental input frequency. This term is being replaced by a new term, PARD (Periodic and Random Deviation). PARD includes all hum, noise, and spikes that occur in the output while the load is held constant.

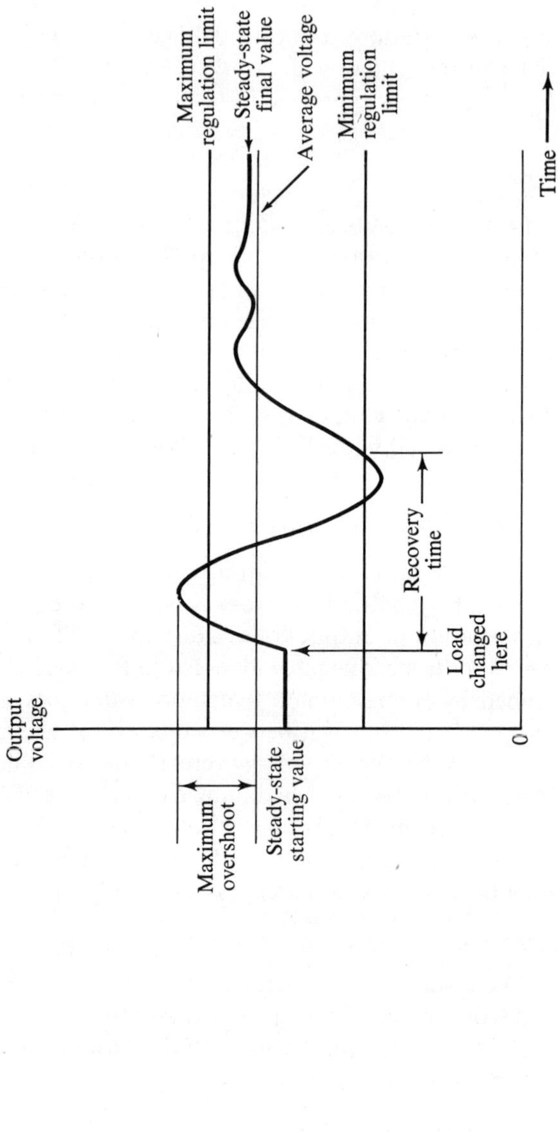

FIG. 6-13. Dynamic-response curve for typical power supply.

Ripple or PARD is expressed as an rms value or as a percent of average output voltage at 50 percent of rated load current.

Thermal-Regulation or Temperature Coefficient

These terms refer to the variation in output voltage when the load current is held at 50 percent of rated value and the ambient temperature surrounding the power supply is varied within the specified limits. Thermal regulation is expressed as percent per-degree Centigrade.

Stability or Drift

These terms refer to the variation in output voltage, after rated warm-up period, while the unit is operated at 50 percent of load at room temperature and nominal input voltage. Stability is measured over a 24-hour period and expressed in percent of output voltage or as an absolute value.

Internal Impedance

Internal (output) impedance is the impedance presented at the load terminals of the power supply. This impedance is calculated as the ratio of the change in output voltage to change in output current when the output current is changed from 50 percent of the rated value to approximately 75 percent of the rated value.

Short-Circuit Proof

A short-circuit proof indicates that the power supply will not be damaged if a short-circuit is placed across the output terminals.

Voltage-Current Crossover

This term refers to that feature of a power supply which changes the regulation mode from constant-voltage-regulation to constant-current-regulation as determined by preset limits. The constant-voltage- and constant-current-settings are independently adjustable within the preset limits.

6-10 VOLTAGE-REGULATION OF D-C GENERATORS

In areas where an a-c supply is not available, the d-c power required to operate electronic equipment may be taken directly from portable, engine-driven, d-c generators. Power for portable electronic field equipment is often obtained in this manner. Generators of direct current that are driven by variable-speed internal-combustion engines require automatic voltage-regulation circuits that will compensate for output voltage-variations caused by changes in load or in engine-speed.

Figure 6-14 shows the basic regulation system for a self-excited d-c generator.

A reference voltage is used to establish a bias current I_b in the control transistor. As long as the output or reference voltage does not change, the system is balanced. When the output voltage changes, the base current also changes. Since the transistor is a current-amplifying device, the field current will change by an amount βI_b, where β is the current gain of the transistor. The system is designed so that a change in field current is in the direction that will cause a return of the output voltage to the balanced condition.

Referring now to the a-c equivalent circuit in Fig. 6-14(b), an error voltage V_e is seen to exist whenever the output of the generator deviates from the steady-state value. A suddenly-applied load could cause this change in output. The a-c error voltage is the difference between the reference voltage and the instantaneous output voltage of the generator. This type of system is called a *proportional control* system because the base current fed back through the transistor to control the field current is proportional to the error voltage.

To determine the effectiveness of the regulation system, the expression for the change in output voltage as a function of load-current change with a transistor in the feedback circuit will be derived. This expression will then be compared with the change in output voltage without any transistor in the feedback loop. The amount of improvement in regulation can then be obtained. The block diagram for the system is shown in Fig. 6-14(c). All frequency terms have been omitted, as only the steady-state errors are of interest.

The equations for the transistor base current at no-load and full-load are

$$I_{b(NL)} = \frac{E_{NL} - E_R}{R_1}$$

and

$$I_{b(FL)} = \frac{E_{FL} - E_R}{R_1} \tag{6-24}$$

The generator field current is a function of the transistor base current and current gain or

$$I_f = \beta I_b \tag{6-25}$$

The output voltage of the generator is a function of the field current and the generator constant K_g or

$$E_{NL} = I_{f(NL)} K_g$$

and

$$E_{FL} = I_{f(FL)} K_g \tag{6-26}$$

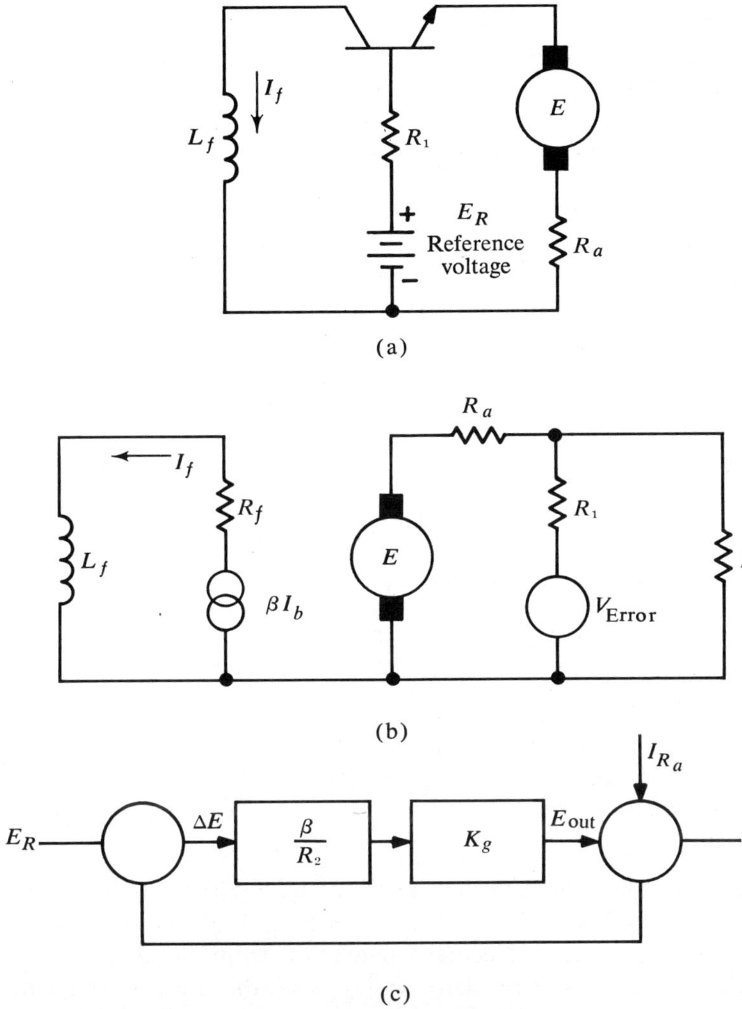

FIG. 6-14. (a) Self-excited d-c generator with transistorized regulator; (b) Circuit equivalent to that in (a); (c) Block diagram.

The equations for the generator output voltage at no-load and full-load, respectively, are

$$E_{NL} = \frac{E_{NL} - E_R}{R_1} \beta K_g$$

and

$$E_{FL} = \frac{E_{FL} - E_R}{R_1} \beta K_g - I_a R_a \tag{6-27}$$

where I_aR_a is the voltage drop across the internal resistance of the generator. For the regulated system the difference between the no-load and full-load voltage is written

$$E_{NL} - E_{FL} = (E_{NL} - E_{R})\frac{\beta K_g}{R_1} - \left[(E_{FL} - E_{R})\frac{\beta K_g}{R_1} - I_aR_a\right] \quad \textbf{(6-28)}$$

Rearranging Eq. 6-28,

$$E_{NL} - E_{FL} = \frac{\beta K_g}{R_1}(E_{NL} - E_{FL}) + I_aR_a \quad \textbf{(6-29)}$$

Equation 6-29 reduces to

$$(E_{NL} - E_{FL})\left(1 - \frac{\beta K_g}{R_1}\right) = I_aR_a \quad \textbf{(6-30)}$$

from which

$$E_{NL} - E_{FL} = \frac{I_aR_aR_1}{R_1 - \beta K_g} \quad \textbf{(6-31)}$$

If $R_1 \ll \beta K_g$, then Eq. 6-31 reduces to

$$E_{NL} - E_{FL} = -I_aR_a\left(\frac{R_1}{\beta K_g}\right) \quad \textbf{(6-32)}$$

Without feedback the difference between the no-load and the full-load voltages is I_aR_a, or the internal voltage drop.

The difference between the no-load and the full-load voltages is now reduced by a factor dependent upon the generator constant and gain of the transistor.

Example 6-6 The following data are available for the circuit of Fig. 6-14: $R_a = 1\ \Omega$; $E_{NL} = 200$ v; $I_{FL} = 20$ A; $K_g = 50$ v/A; β transistor $= 100$; and $R_1 = 1000\ \Omega$. Find the percent of regulation from no-load to full-load: (a) without the transistor-regulator and (b) with the transistor-regulator

$$\text{Percent reg} = \frac{E_{NL} - E_{FL}}{E_{NL}} \times 100$$

Without the regulator

$$\text{Percent reg} = (I_aR_a/E_{NL})(100)$$

or

$$\text{Percent reg} = \frac{I_aR_a}{200} = \frac{20}{200} = 10 \text{ percent}$$

With the transistor-regulator

$$E_{NL} - E_{FL} = \frac{20(1)(1000)}{100 \times 50} = 4 \text{ v} \qquad \text{(From Eq. 6-32)}$$

$$\text{Percent reg} = \frac{4}{200} = 2 \text{ percent}$$

Thus, the transistor-regulator has provided a 5-to-1 improvement in regulation.

6-11 THE AMPLIDYNE

A device that is very useful for controlling large amounts of d-c power is the amplidyne, shown schematically in Fig. 6-15.

The amplidyne is a rotating d-c amplifier which, with small amounts of control power, can control high-power levels at a load. The device is driven by a prime mover, usually an a-c or a d-c motor. Current in the control field produces the control flux Φ_2. When winding 1, shown passing under the pole face in Fig. 6-15(b), cuts this flux, a voltage is induced in this winding. A set of brushes is so located as to short-circuit the winding that passes under the pole faces. This short-circuit produces a high current through this winding, which in turn produces the flux Φ_3 at right angles to flux Φ_2. Winding 2, connected to the load through a separate set of brushes, is located at 90° from winding 1 on the armature shaft. A voltage is induced in this winding as it cuts the flux Φ_3 created by winding 1. This voltage is used to deliver power to the d-c load.

The current through winding 2 may create a flux that tends to oppose the control flux Φ_2 and reduce the power output of the device. To counteract any opposing flux, a compensating field winding is placed in series with winding 2 and the load. Current in this compensating winding produces the flux Φ_5 that aids the control field. The output is maintained.

The power gain of the amplidyne is very high. Control winding-power of 1/4 to 1/2 W can control load power as high as 2.5 kw. The power gain can, therefore, be of the order of 10,000-to-1, or higher.

The amplidyne has a very short time-constant, and fast response can be obtained from amplidyne systems. An equivalent circuit for the amplidyne is given in Fig. 6-16. The device may be represented as a two-stage d-c generator. If each generator has a gain K_g of 20 v/A, then

$$E_L = I_{f_2} K_g$$

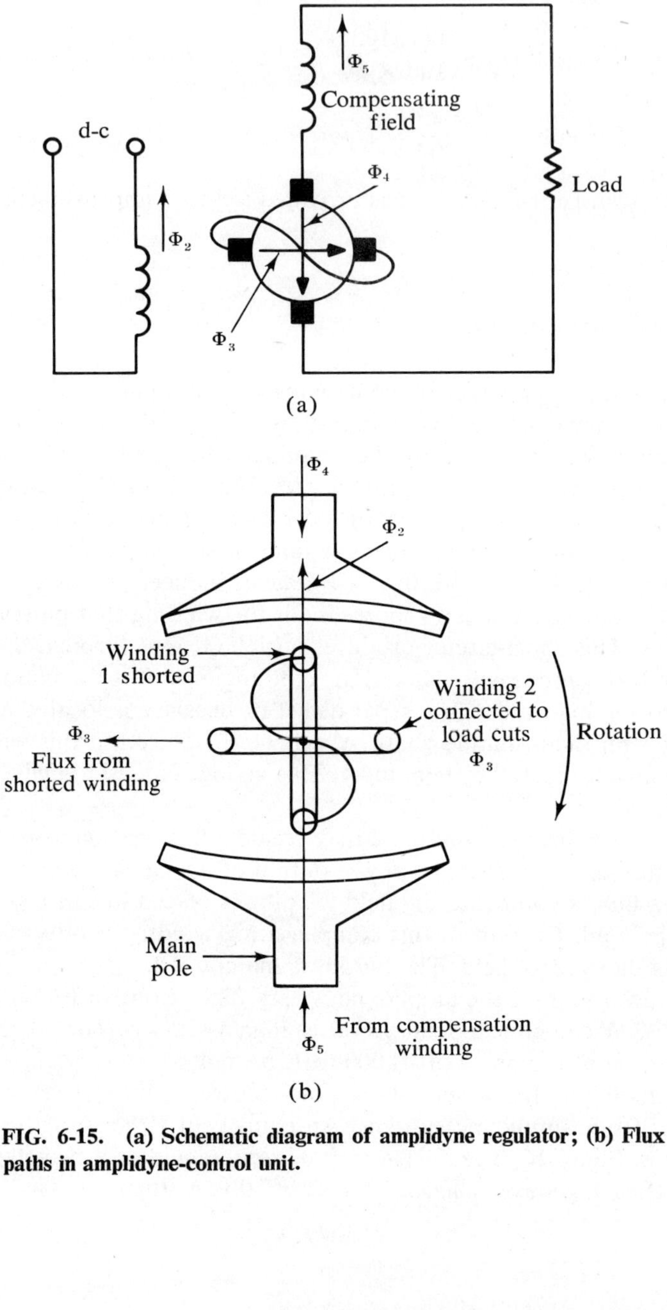

FIG. 6-15. (a) Schematic diagram of amplidyne regulator; (b) Flux paths in amplidyne-control unit.

If

$$I_{f_2} = \frac{E_{g1}}{R_{f_2}}$$

then

$$E_L = \frac{E_{g1}}{R_{f_2}} K_g$$

and

$$\frac{E_L}{E_{g_1}} = \frac{K_g}{R_{f_2}}$$

For both generators cascaded,

$$E_{g_1} = \frac{E_f}{R_{f_1}} K_g$$

and

$$\frac{E_L}{E_f} = \frac{K_{g1} K_{g2}}{R_{f_1} R_{f_2}}$$

The overall gain, therefore, is a function of the product of the individual gains.

In Fig. 6-15(b), flux Φ_2 and winding 1 represent generator 1, and flux Φ_3

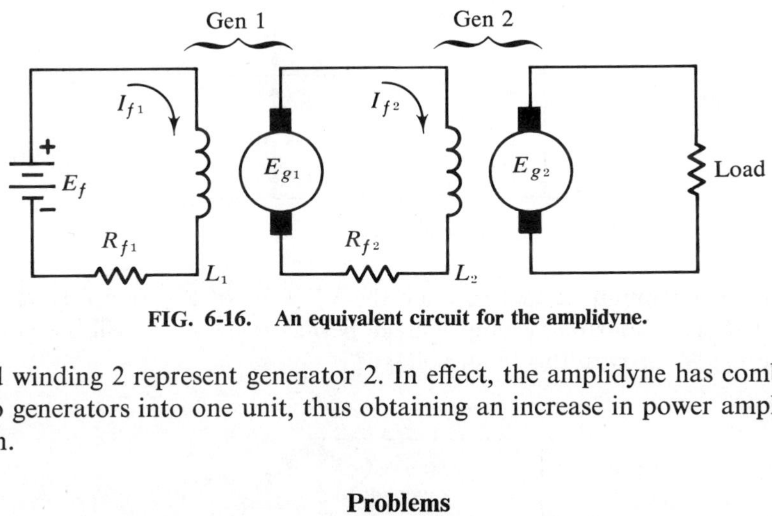

FIG. 6-16. An equivalent circuit for the amplidyne.

and winding 2 represent generator 2. In effect, the amplidyne has combined two generators into one unit, thus obtaining an increase in power amplification.

Problems

1. Using a zener diode with the characteristics shown in Fig. 6-3, design a simple regulator that will maintain 75 v across a 1000-Ω load as the input voltage varies from 80 to 120 v. What will be the maximum current in the zener diode?
2. Over what input-voltage range will the regulation circuit of Problem 1 regulate when diode current is limited to 0.5 A?

3. Using two zener diodes, one rated at 100 v, 0.3 A and the second at 50 v, 0.2 A, design a cascade-regulation circuit that will maintain 50 v across a 200-Ω load, minimum input-voltage of 110 v. What will be the maximum allowable input voltage to this circuit?
4. It is desired to maintain the voltage across a 200-Ω resistor at 50 v. A 50-v, 0.5-A zener is available, and the minimum input voltage expected at the source is 75 v. No more than 16 W may be wasted in a series resistor. Design the zener circuit that will meet these requirements. What will the maximum input voltage be?
5. If the no-limit is placed on the watts lost in Problem 4, over what input-voltage range can the circuit regulate?
6. In Fig. P6-6, what percentage change in voltage will appear across R_L if the input voltage rises by 6 percent?

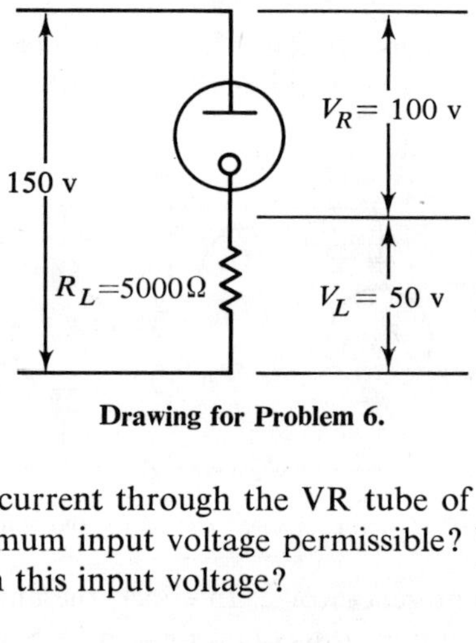

Drawing for Problem 6.

7. If the maximum current through the VR tube of Problem 6 is 40 mA, what is the maximum input voltage permissible? What will the voltage across R_L be with this input voltage?

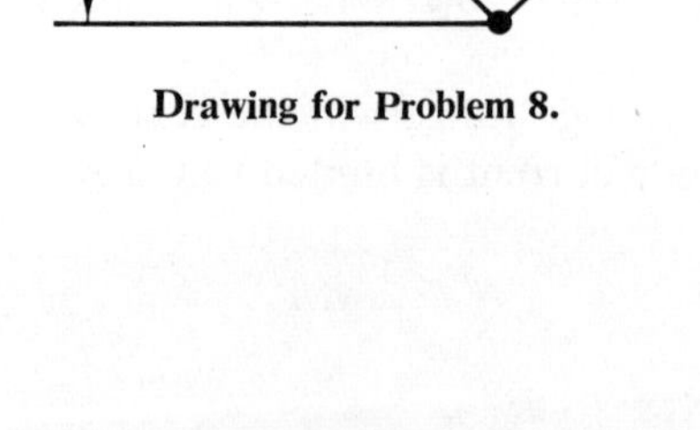

Drawing for Problem 8.

8. Refer to Fig. P6-8. Plot a curve of the input voltage from zero to 150 v vs E_{AB} for the zener bridge circuit shown. Indicate the point at which the zener fires and the voltage at which E_{AB} is zero.
9. The 50-v zener of Problem 8 is rated at 5 mA for normal operation and is to be used in a circuit which has an operating voltage of 70 v. At this voltage V_{AB} is to be zero. Design a circuit to meet these specifications.
10. Over what range of input voltage will the zener-regulating circuit of Fig. 6-5 maintain 30 v across the 2000-Ω resistor? Maximum zener current is not to exceed 60 mA; $R_s = 100\ \Omega$.
11. What will be the maximum power loss in the regulation circuit of Problem 10?
12. Referring to the circuit of Fig. 6-6(b), $g_m = 50 \times 10^{-4}$; $R_1 = 10$ k; $R_2 = 2$ k; $R_L = 2$ k; $R_s = 10$ k; and $R_P = 10$ k. What will be the voltage regulation constant of the circuit under these conditions?
13. The voltage-regulation constant of Problem 12 is to be improved to 0.035. If the R_s is changed to improve K_V, what will the new value of R_s be? If R_L is to be changed, what must its new value be?
14. For the series-regulation circuit of Fig. 6-8, $\beta = 50$; $R_b = 10$; $R_c = 1000\ \Omega$; and $R_e = 20\ \Omega$. Find the voltage-regulation constant for the circuit under these conditions.
15. What will the new regulation constant be if β is increased to 100?
16. The output voltage of a power supply varies from 199 to 201 v as the input varies from 200 to 210 v. What is the percentage regulation?
17. The output voltage of the power supply of Problem 16 varies from 201 to 198 v as the load varies from zero to rated value. What is the percentage of load regulation?
18. What is the total regulation for the power supply of: (a) Problem 16? and (b) Problem 17?
19. The following data are available for the generator circuit of Fig. 6-14; $R_a = 0.3\ \Omega$; $R_1 = 500\ \Omega$; $\beta = 100$; and $K_g = 50$ vA. The generator is rated at 200 v and 50 A. What will the change be in voltage from no-load to full-load, (a) with the series-regulation circuit? and (b) without the series-regulation circuit?
20. What will be the percentage of regulation of the generator in Problem 19 when β is decreased to 50?
21. If the generator of Problem 1 is to have a regulation of 1 percent, what must the value of β for the transistor be?
22. The following data are available for the circuit of Fig. 6-14: $R_a = 0.2\ \Omega$; $E_{NL} = 200$ v; $I_{FL} = 50$ A; $K_g = 60$ vA; $\beta = 80$; and $R_1 = 1000$. Find the percentage of regulation of this generator from no-load to full-load,

(a) with the transistor-regulation circuit and (b) without the transistor-regulation circuit.

Questions

1. Name the necessary elements for a feedback-control system.
2. Define "recovery time" of a voltage-regulation system.
3. Why is a reference voltage needed for a regulated power supply?
4. Define "overshoot" in a voltage-regulation system.
5. Explain how a VR tube can be made to amplify changes in input voltage.
6. Why is it desirable to amplify the changes in the regulated voltage of a power supply?
7. Draw a circuit showing how a VR tube may be used to amplify the changes in output voltage of a power supply.
8. Draw a circuit using a zener diode as an error-amplifier.
9. Draw a circuit that uses the voltage across a zener diode as a reference voltage.
10. Why is a zener bridge circuit, such as that shown in Fig. 6-4, a more sensitive error-detector than the VR circuit shown in Fig. 6-2?
11. What factors determine the accuracy of a simple zener diode voltage-regulation circuit?
12. What component of the circuit in Fig. 6-5 must be changed to extend the range of input voltage over which the load voltage will be maintained? How must it be changed?
13. Describe the operation of a series-regulation circuit.
14. Describe the operation of a parallel-regulation circuit.
15. Describe the operation of a line-regulation circuit.
16. Describe the operation of a line-load-regulation circuit.
17. What are the prime factors that determine the accuracy of the shunt regulator shown in Fig. 6-7?
18. Define "voltage-regulation constant."
19. How will the accuracy of the shunt-regulator shown in Fig. 6-7 be affected if the emitter resistor R_e is increased?
20. What are the disadvantages of the shunt-regulator, the series-regulator?
21. Which transistor in the series-regulation circuit of Fig. 6-8 must have the greatest power rating? Why?
22. How will the voltage-regulation constant of the series-regulator shown in Fig. 6-8 be affected if β for transistor 1 is increased?
23. Why is it necessary to compensate transistorized regulators?
24. Sketch a circuit that may be used as a transistorized series-current-regulator.

25. Why is current-regulation desirable in power supplies?
26. Define the total regulation of a power supply.
27. Define dynamic-load-regulation.
28. What factors contribute to the internal impedance of a power supply?
29. Define proportional control.
30. Describe the operation of an amplidyne.
31. Why are the armature windings on an amplidyne placed 90° apart?
32. Why is the armature winding of one amplidyne shorted by brushes?

Answers to Odd-numbered Problems

Chapter 1

1-1. Average values, Sq. wave 50 V, Sawtooth 50 V, Triangle 50 V
Rms values, Sq. wave 50 V, Sawtooth 57.7 V, Triangle 57.7 V
1-3. 11.1 A
1-5. (a) 800 Hz (b) 60 Hz (c) 120 Hz (d) 120 Hz (e) Half-wave, single-phase rectifier
1-7. 40°, 90°
1-9. .3 A
1-11. 165 V
1-13. 61.5°, 1 peak = 2.2 A, 1 avg = .76 A
1-15. $C = 2.1\ \mu f$ $E_{rms} = 101.8$ V $E_{a\text{-}c} = 17.2$ V $\phi_c = 63°$ 1 peak = .612 A
$P = 5.16$ W
1-17. $E = 67.5$ V (peak-to-peak)
1-19. 14.4 W
1-21. Full-wave, $E_{avg} = 157$ V, $E_{rms} = 157.1$ V, $\gamma = 4.83$ percent, $\phi_c = 34.4°$, $Tc = .00157$ sec
Half-wave, $E_{avg} = 146$ V, $E_{rms} = 147$ V, $\gamma = 9.65$ percent, $\phi_c = 46.5°$, $Tc = .00215$ sec
1-23. $L = .5$ H, $\gamma = 40$ percent $L = .7$ H, $\gamma = 37.4$ percent
1-25. $C = 2, 100\ \mu f$
1-27. $C = 1, 100\ \mu f$
1-29. $E_{rms} = 120$ V, $E_{avg} = 107.8$ V, $E_{a\text{-}c} = 53$ V, $P = 14.4$ W, 01 V = 338 V

Chapter 2

2-1. $\phi_c = 120°$, $E_{avg} = 181$ V, $E_{rms} = 182$ V, $E_{a\text{-}c} = 31$ V, $P = 169.5$ W
2-3. 248 VA
2-5. 14.85 A
2-7. 25.5 A
2-9. $I_{avg} = 14.5$ A, PIV = 605 V, VAs = 18,750, $I_{rms} = 14.7$ A, Ripple = 85 V
2-11. $E_{max} = 204$ V, Ripple = 16.6 V, $P_{avg} = 800$ W

2-13. I/diode = 6.66 A, VAs = 3560, V_{sec} = 126 V, I_{rms} = 20.1 A, Ripple = 11.8 V
2-15. VA = 15,100, V_{sec} = 208 V (line-to-neutral)
2-17. E = 56.4 V (peak-to-peak)
2-19. .785 μf
2-21. 38 kW
2-23. 405 V

Chapter 3

3-1. 50 H
3-3. 1000 Wb @ .5 A, 4000 Wb @ 2 A, L = .25 H
3-5. ϕ_f = 31° @ 60,000 lines ϕ_f = 60° @ 35,000 lines
3-7. 160 W
3-9. 1125
3-11. With positive feedback
W_{load} = 36.9, $W_{control}$ = 53, Kc = 10, Power gain = 12.5, Tc = .052 s
With negative feedback
W_{load} = 8.1, $W_{control}$ = 53, Kc = 3.33, Power gain = 1.42, Tc = .0058
3-13. Tc = .3 s, W = 20,000

Chapter 4

4-1. 1820 Ω
4-3. 145.5 to 168.8 V
4-5. 434 to 476 Ω
4-7. E_{in} can vary from 110 to 200 V
4-9. Eg = −2.3 V
4-11. $R_1 = R_2$ = 3770 Ω
4-13. Eg = −2, ϕ_f = 9.5°, Eg = −4, ϕ_f = 30°, Eg = −6, ϕ_f = 90°
4-15. R = 1 kΩ, L = 3.2 H
4-17. Will not fire if the phase shift between the line and grid is 90°.

Chapter 5

5-1. E_{avg} varies from $E_{max}/2\pi$ to zero
5-3. 48°
5-5. 17 W
5-7. 160° to 2°
5-9. ≈ 90°
5-11. 8 percent
5-13. $\phi = 0$, E = 220 V
ϕ = 60°, E = 159 V
ϕ = 120°, E = 151 V

5-15. $\phi = 0, E = 220$ V
$\phi = 60°, E = 159$ V
$\phi = 120°, E = 151$ V
$\phi = 180°, E = 0$

Chapter 6

6-1. 600 ma
6-3. $V_{max} = 170$ V
6-5. 75 V to 125 V
6-7. Maximum input voltage is 300 V. Voltage across the load is 200 V.
6-9. $R_1 = 10$ k, $R_2 = 4$ k, $R_3 = 4$ k, $I_1 = 5$ ma
6-11. .565 W
6-13. $R_s = 19.2$ k $R_L = 545\ \Omega$
6-15. .0195
6-17. -1.495 percent or -3 V
6-19. 15 V without regulation. 1.5 V with regulation
6-21. $\beta = 75$

Index